Farming with the Environmer

This book examines, discusses and shares over 30 years' worth of research from the Allerton Project, a research and demonstration farm in the UK which has been carrying out applied interdisciplinary research to explore and explain the need to adapt the management of farmland for environmental protection and to provide public benefits.

Designed to provide guidance, feedback and recommendations to farmers, practitioners and policymakers, the Game & Wildlife Conservation Trust's Allerton Project is an exceptionally well-documented case study of lowland agricultural land management which has the purpose of meeting multiple objectives. This book draws on the wealth of knowledge built over the past 30 years and unveils and clarifies the complexity of a number of topical debates about current land and wildlife management at a range of spatial scales, explores the underlying historical context and provides some important pointers to future directions of travel. Topics include soil health and management, farmland ecology, development of management practices to enhance biodiversity, natural flood management, water quality and aquatic ecology. Most importantly, the book demonstrates how the findings from this project relate to agricultural and conservation policy more broadly as well as how they are applicable to similar projects throughout Europe.

This book will be of great interest to professionals working in agricultural land management and conservation, as well as researchers and students of agri-environmental studies and agricultural policy.

Chris Stoate is an agro-ecologist specialising in identifying synergies between agricultural and environmental objectives within lowland farming systems at a range of scales. He is Head of Research at the Allerton Project research and demonstration farm and has worked for the project since it was established by the Game & Wildlife Conservation Trust in 1992. Chris is an Honorary Professor with the University of Nottingham and a farmer in his own right, and has worked in southern Europe and West Africa as well as in the UK.

Earthscan Food and Agriculture

For more information about this series, please visit: www.routledge.com/books/series/ECEFA/

Farming with the Environment

Thirty Years of Allerton
Project Research

Chris Stoate

Routledge
Taylor & Francis Group
LONDON AND NEW YORK

earthscan
from Routledge

Cover image: The Allerton Project research and demonstration farm
at Loddington

First published 2023
by Routledge
4 Park Square, Milton Park, Abingdon, Oxon OX14 4RN

and by Routledge
605 Third Avenue, New York, NY 10158

Routledge is an imprint of the Taylor & Francis Group, an informa business

British Library Cataloguing-in-Publication Data
A catalogue record for this book is available from the British Library

Library of Congress Cataloging-in-Publication Data
A catalog record has been requested for this book

ISBN: 978-0-367-74900-2 (hbk)
ISBN: 978-0-367-74897-5 (pbk)
ISBN: 978-1-003-16013-7 (ebk)

DOI: 10.4324/9781003160137

Typeset in Times New Roman
by codeMantra

Contents

Illustrations

Figures

Tables

Boxes

Preface

On 18 February 1992, I drove from my then home in Hampshire to Loddington in Leicestershire to conduct a night-time spotlight count of brown hares. That was the start of 30 years of data collection on what was to become the GWCT Allerton Project research and demonstration farm. This was the result of the extraordinary generosity of Lord and Lady Allerton who previously owned the farm, and the foresight of the first chairman of the Allerton Research and Educational Trust, Philip Grimes.

With Dr. Nigel Boatman as our first project director, our initial research focus was very much on farmland ecology, and one of our first practical and policy achievements was establishing the evidence base for wild bird seed mixtures as an agri-environment scheme option for adoption across the country. Through the 1990s, I also contributed to research in the Alentejo region of southern Portugal where my thinking was influenced by the montado farming system, which is one of the classic examples of agroforestry, but more broadly is an exemplar of multifunctional resource use on farmland.

The concept of ecosystem services began to be popularised during this period, and it became clear that our research needed to expand to cover topics such as water quality and catchment management, soil management and pollinators as well as broader aspects of aquatic and terrestrial biodiversity.

Nigel and I worked with other European partners on a contract for the European Commission's Directorate-General for Environment, reviewing the environmental impacts of changes in farming practices across the EU. One of the key messages from this work was the importance of soil as being fundamental to a wide range of objectives for agricultural land. We were fortunate in appointing Dr. Alastair Leake as our second project director, bringing to the project a deep understanding of integrated farming and soil management. Alastair was instrumental in initiating some of the projects referred to in Chapter 3.

Over the same period, I carried out research in an area in West Africa where farmland and associated coastal fisheries had become severely degraded as a result of runoff and soil erosion. The research linked soil management with the conservation of migratory birds through the incorporation of certain tree species into farming systems. It also highlighted the international integration of wide-ranging issues, including food production and the conservation of some of the migratory bird species present at Loddington during the summer.

From the start of the project in 1992, Ecologist John Szczur has brought an exceptional knowledge of plant, invertebrate and vertebrate ecology to the project, while Farm Manager Phil Jarvis played a key role in establishing experimental plots and recording crop yield and economic data that became core to our research in recent years.

However, the breadth of our research has only been possible through collaboration with researchers from other organisations and universities. Multiple collaborations over the years have brought specialist expertise that has enabled us to cover topics as diverse as aquatic ecology, water quality, flood risk management, soil management and biology, carbon sequestration, ruminant nutrition, greenhouse gas emissions, agroforestry, game management, farmland ecology, pollinators and crop pest predators. This interdisciplinary approach has also extended into the social sciences and economics, recognising the multiple influences on farmers' decision-making. It has been a real privilege to work alongside and learn from these research partners and numerous colleagues over the years. There have been too many to name here, but their contributions are acknowledged individually by the citations throughout this book. Their collective contribution has been enormous.

Active involvement in farming, both personally on my own land and as an organisation on the farm at Loddington, ensures that our work is kept practically grounded, while always pushing at the boundaries. It has also enabled us to work with other members of the local farming community by adopting a participatory approach to our research. I am grateful to those farmers for their collaboration.

The learning from this broad and integrated knowledge base has also enabled Allerton Project staff to play an influential role in various agricultural, environmental and policy arenas, ensuring that our research findings have practical and policy impact. In 2011, I contributed to the UK's National Ecosystem Assessment which underpins current plans for valuing natural capital and the associated ecosystem services. This provides the foundation for the current aspiration to develop and act on environmental land management plans at farm and landscape scales following the UK's departure from the European Union in 2021.

Publication of this book comes during a period of considerable change in the way food is produced and agricultural land is managed, not only in the UK but across much of Europe, North America and other parts of the globe. This is an appropriate moment to share our learning to inform this process.

1 Introduction

Agricultural land occupies around half of the European land area, so its sustainable management is central to meeting our objectives for food security, wildlife conservation, clean water, flood prevention, climate change mitigation, recreation and others. The interactions between these various objectives for the farmed environment over the past century are well documented, and steps have been taken to redress the balance between them within the constraints of the existing policy framework (Stoate et al., 2001, 2009). We are now entering a period of major change for European agriculture as attempts are made to rebalance the need for food production with the protection of the environment on which food production and much else depend.

The policy context

This interdependence of these multiple objectives was officially recognised internationally in the United Nations Millennium Ecosystem Assessment (MA, 2005) which assessed the consequences of ecosystem change for human well-being. The MA also explored the options to restore, conserve or enhance the sustainable management and use of ecosystems. In Europe, the European Union published a 'Biodiversity Strategy to 2020' (EU, 2011) to meet targets for the Convention of Biodiversity (UNEP, 2010), and an accompanying ecosystem assessment (European Environment Agency, 2015). The UK had previously published its own ecosystem assessment (UK NEA, 2011) with a chapter relating specifically to ecosystem services associated with agricultural land which represents around half of the UK land area (Firbank et al., 2011).

While some north-western European countries had implemented early agri-environment schemes in the late 1980s, since 1992, the European Union has required and supported the adoption of agri-environment schemes to meet a range of environmental objectives, with uptake varying in scope and extent across all the member states. The USA also adopts voluntary incentive-based schemes to complement regulatory programmes and cross compliance (Claassen et al., 2014). Norway and Switzerland adopted parallel

DOI: 10.4324/9781003160137-1

schemes to those in the EU. While there have been specific successes of some of these schemes where they are most targeted and introduce the greatest landscape change (e.g. Walker et al., 2018), overall biodiversity benefits have been described as 'moderate' (Batáry et al., 2015).

The EU's Common Agricultural Policy is being reviewed to deliver on the EU Green Deal and associated Biodiversity and 'Farm to Fork' Strategies (EC, 2020a, 2020b). Together, these strategies aim to address agriculture's contribution to climate change in line with the United Nations 2015 Paris Agreement on climate change, for example by improving resource use efficiency and encouraging carbon sequestration, and to enhance biodiversity through measures such as Integrated Pest Management and increasing the area of semi-natural habitats. This is intended to be achieved while simultaneously improving the economic competitiveness and sustainability of agricultural production, ensuring the continuing secure supply of abundant and nutritious food, as well as other agricultural products.

The UK left the European Union on 31 January 2020 and is adopting a similar but independent strategy to its former EU partners. At the core of this is an Environmental Land Management scheme with similar objectives to those of the EU Green Deal and a range of levels of ambition. Historically, the UK's spending on agri-environment schemes has been higher than that of most other EU countries, and the country also has a strong evidence base, including long-term monitoring, development of agri-environmental management practices, and evaluation of the management options that are implemented.

These national, international and global policy changes make multidisciplinary research projects on agricultural ecosystems especially important, particularly where they are long term and operating at a range of scales. Initiatives with a strong biodiversity element such as the Schorfheide-Chorin research project in northern Germany (Flade et al., 2006) and long-running research in the Zone Atelier and Val de Sèvre of western France (Gaba & Bretagnolle, 2020) provide a strong evidence base but are associated with large areas that are designated as being of existing high nature value in the form of a Biosphere Reserve and Natura 2000 site, respectively. In addition to these important long-term interdisciplinary research projects, there is a need for equivalent work in areas that are more representative of wider agricultural systems.

The Allerton Project

The wide-ranging issues associated with ecosystem services in an agricultural setting have been investigated through a long series of research projects by the Allerton Project, a research and demonstration farm in the English East Midlands. The East Midlands is a predominantly rural region in which agriculture is a major land use and has a wide range of soil types from clays to lime-rich loam. Arable crops predominate, mainly comprising wheat (53%), oilseed rape (22%) and barley (17%), with vegetables (4%) and sugar beet (3%) on lighter land to the east (Defra, 2018). Livestock systems are mainly sheep and cattle at a ratio of about 2.5 to 1.

The 333 ha Allerton Project farm is located in the upper, western part of the Welland river basin 105–180 m above sea level, with undulating topography and relatively low-grade clay soils. As such, it is representative of about a third of lowland England (Brown et al., 2006). In terms of soil type, topography and farming system, it is also representative of a large part of lowland continental Europe. Figure 1.1 shows the distribution across Europe of areas where our research into arable soil management and grass leys would also be applicable.

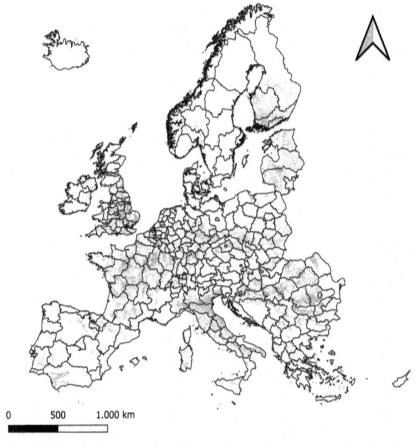

Figure 1.1 Map of Europe showing the distribution of areas that share the same conditions for crop establishment as the Allerton Project farm at Loddington, Leicestershire (van Delden, 2021)

The farm at Loddington is the former home of Lord and Lady Allerton who bought the estate in 1934. The executors of their will set up the Allerton Research and Educational Trust in April 1992 to own and manage the farm. The Trust's objectives were to conduct research and disseminate research results in order to advance public education in different farming methods and their effects on the environment and wildlife. These activities were supported by the Game & Wildlife Conservation Trust (then The Game Conservancy Trust) which had a long history of research into agricultural ecosystems. The project formally became the Game & Wildlife Conservation Trust's 'Allerton Project' in 2006.

The farm has been managed as a commercial business and comprises 253 ha of arable land, 29 ha of pasture and 18 ha of mature woodland (Figure 1.2). Crop production has varied annually with cropped land area and the farm's yields but profitability has been comparable to that of other local farms. The crop rotation has evolved over the 30 years of the project in response to global markets and local pressures such as weed burdens and changes in weather. The major crops have been winter wheat, oilseed rape and field beans, although more niche crops such as hemp, flax and linseed have also been grown. Winter barley was grown in the early years, and then brought back into the rotation in more recent years after a period of absence. Grass leys have been incorporated into the rotation from 2017, with more diverse herbal leys being trialled from 2019. The area of land taken out of production to meet EU set-aside requirements in 1992 was 27 ha, rising to a maximum of 43 ha in 1994 and reducing to zero in 2007 when the requirement ended. The area of land devoted to habitat management and to meet other environmental objectives has ranged from 10 to 15% of the agricultural land area through most of the period.

For nearly two decades, the farm business was managed as a joint venture with a neighbouring farm, benefiting both businesses by spreading fixed costs. Labour and machinery were pooled and deployed across the combined area. This enabled the farm to invest in the necessary equipment to adopt a reduced tillage system, further reducing crop establishment costs and delivering environmental benefits. Contract work on other local farms also contributes to the farm's income. However, a looser working relationship has been adopted more recently as research contracts have been absorbed more closely into the farm business and made different demands in terms of management practices, labour and equipment.

For the first 20 years, the pasture was grazed by a flock of 280 mule ewes, but since 2011 the grazing has been let to a neighbouring farm which specialises in sheep. This released late winter labour for hedge laying and woodland management within the agri-environment and farm woodland agreements, removing the need to hire contractors for this work. More recently, an arrangement with a local contractor consists of an exchange of usable timber for woodchip produced from smaller branches to heat the project's visitor centre and offices.

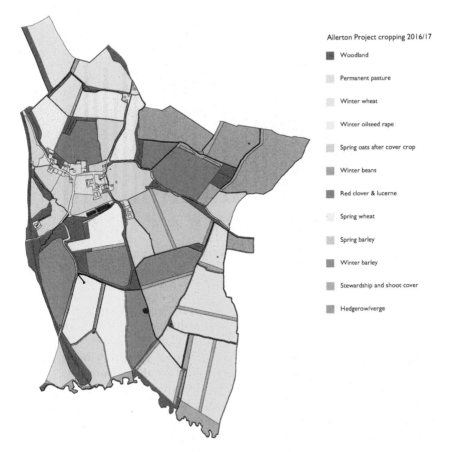

- Woodland
- Permanent pasture
- Winter wheat
- Winter oilseed rape
- Spring oats after cover crop
- Winter beans
- Red clover & lucerne
- Spring wheat
- Spring barley
- Winter barley
- Stewardship and shoot cover
- Hedgerow/verge

Figure 1.2 Map of the Allerton Project farm at Loddington showing the distribution of arable crops, grassland and woodland in 2017

Along with many other farms across lowland England and continental Europe, the farm is on marginal land and crop yields vary considerably from year to year with a consequent impact on overall food production from the farm (Figure 1.3). In recent years, food production has been compromised by drought, and more substantially by intense storms and waterlogged, compacted ground. Climate change impacts therefore represent an increasing challenge for the business.

The Common Agricultural Policy's Single Farm Payment has been critically important to the farm business and without it, and in common with others, the farm would not have been profitable for many years. Agri-environment scheme agreements have provided a useful source of income to compensate for the land taken out of production and the necessary management to create habitats on that land, as well as paying for routine stewardship

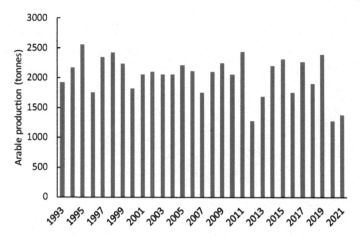

Figure 1.3 Overall arable food production (1992–2021) using wheat as a proxy for all crops

of hedgerows and field margins. However, minimising costs in the joint partnership through reduced overheads, maximising yields through good timeliness and agronomy, marketing crops with 'Conservation Grade' quality premiums, and forward selling at times, have all been very important in achieving our main priorities of resilient and profitable farming.

More recently, as research in the cropped area has increased, this has brought an income stream to the business. With the UK's departure from the EU and a transition away from area payments, the Allerton Project's farm business is having to adapt to more economically challenging circumstances. This will include participation in the UK government's Environmental Land Management (ELM) scheme which makes 'public payments for public goods' in the form of ecosystem services. The Allerton Project's research presented in this book will continue to play a key role in informing this process at the farm scale, nationally, and beyond.

The research carried out over the 30 years of the project has covered a wide range of agri-environmental issues. Initially, this comprised various aspects of farmland ecology (especially birds), including the pioneering development of management practices that are now widely adopted as part of national agri-environmental schemes. Other topics have included a range of issues associated with soil management, aquatic ecology, catchment management, agroforestry, game management, and grass and livestock management to meet multiple objectives. This research has necessarily been carried out at a range of scales, from replicated experimental plots to landscape scale catchments. The research has also been interdisciplinary (Stoate, 2008), both within the natural sciences and integrating social sciences. Given that agricultural land is, by definition, managed by people, an understanding of the socio-economic and cultural influences on land management is essential.

Prerequisites for successful long-term ecosystem research include adaptability, a whole system approach, multiple objectives, a balance between time series continuity and adoption of emerging technology, and active stakeholder engagement (Musche et al., 2019). As an independent research centre, the Allerton Project meets all these criteria. This is achieved mainly by accumulating a series of complementary short-term research contracts. Supplementary material comes from small-scale projects and post-graduate student research projects. Citations of journal papers, PhD theses and some other unpublished sources throughout the book pay tribute to the numerous individuals who have contributed to our collective understanding. The aim of this book is to bring together the results of these various projects to create a coherent narrative that will help to inform future policy and practice.

Layout of the book

Historically, we have modified, interacted with and adapted to our environment, a continuing process termed 'cultural ecology' (Stoate, 2001a, 2001b, 2002). The agricultural ecosystem is one which has been substantially modified by people and is shared with multiple other species which are adapted to it (Stoate, 2011). Understanding our historical integration with the land provides a foundation for understanding our current and future relationship with it and is the subject of Chapter 2. Most fundamentally, how we manage the soil and harness the ecological processes within it to provide our food and multiple other ecosystem services is at the heart of land management and discussed in Chapter 3. The recent disconnect between people and the land has resulted in a popular lack of knowledge of the relationship between wildlife and our food – the ecology of food, a topic discussed in Chapter 4. Chapter 5 covers a range of habitats and management practices that have been developed at the Allerton Project to meet multiple objectives for farmland. These management practices are now incorporated in UK agri-environment schemes and widely adopted on farms across other lowland landscapes.

The relationship between land management and water quality and ecology is the subject of Chapter 6. This work ranges in scale from small streams and ponds to the landscape scale, in particular on the Allerton Project's 3,000 ha Water Friendly Farming project. Chapter 7 recognises the complexities associated with many of these issues and highlights the problems associated with polarised views on important issues for land management and wildlife conservation. Chapter 8 returns to the concept of farmers' relationship with the land by exploring the relevant issues from a farmer and farm business perspective, drawing on recent farmer surveys, workshops, participatory research and other social science associated with the Allerton Project. Finally, Chapter 9 looks at how we might draw on this wealth of research to develop plans for more sustainable management of farmland and the production of our food in the future. In this, the UK is embarking on a pioneering approach having left the European Union. None of the issues discussed in this book can be fully considered in isolation and references to other chapters are made where this is important to a comprehensive understanding.

References

Batáry, P., Dicks, L. V., Kleijn, D. & Sutherland, W. J. (2015) The role of agri-environment schemes in conservation and environmental management. *Conservation Biology* 29 (4) 1006–1016.

Brown, C. D., Turner, N. L., Hollis, J. M., Bellamy, P. H., Biggs, J., Williams, P. J., Arnold, D. J., Pepper, T. & Maund, S. (2006) Morphological and physio-chemical properties of British aquatic habitats potentially exposed to pesticides. *Agriculture, Ecosystems and Environment* 113, 307–319.

Claassen, R., Cooper, J., Salvioni, C. & Veronesi, M. (2014) Agri-environmental policies: A Comparison of US and EU experiences. In: Duke, J. M. & Wu, J. (Eds.) *The Oxford Handbook of Land Economics.* Oxford University Press, New York. DOI:10.1093/oxfordhb/9780199763740.013.020

Defra (2018) https://assets.publishing.service.gov.uk/government/uploads/system/uploads/attachment_data/file/697017/regionalstatistics_eastmidlands_04apr18.pdf

European Commission (2020a) Communication from the European Commission, *"A Farm to Fork Strategy for a Fair, Healthy and Environmentally-Friendly Food System."* COM/2020/381 final.

European Commission (2020b) Communication from the European Commission, *"EU Biodiversity Strategy for 2020: Bringing Nature Back Into Our Lives."* COM/2020/380 final.

European Environment Agency (2015) *The European Environment – State and Outlook 2015: An Integrated Assessment of the European Environment.* European Environment Agency, Copenhagen.

European Union (2011) *The EU Biodiversity Strategy to 2020.* Publications Office of the European Union, Luxembourg.

Firbank, L., Bradbury, R., McCracken, D. & Stoate, C. (2011) *Enclosed Farmland.* In: UK National Ecosystem Assessment. *The UK National Ecosystem Assessment Technical Report.* UNEP-WCMC, Cambridge. 197–197–239.

Flade, M., Plachter, H., Schmidt, R. & Werner, A. (2006) *Nature Conservation in Agricultural Ecosystems: Results of the Schorfheide-Chorin Research Project.* Quelle & Meyer Verlag GmbH und Co., Wiebelsheim.

Gaba, S. & Bretagnolle, V. (2020) Designing Multifunctional and Resilient Agricultural Landscapes: Lessons from Long-term Monitoring of Biodiversity and Land Use. In: Hurford, C., Wilson, P. & Storkey, J. (Eds.) *The Changing Status of Arable Habitats in Europe.* Springer Nature, Cham. 203–224.

MA (Millennium Ecosystems Assessment) (2005) *Ecosystems and Human Wellbeing: Synthesis.* Island Press, Washington, DC.

Musche, M., Adamescu, M., Anglestam, P. et al. (2019) Research questions to facilitate the future development of European long-term ecosystem research infrastructures: a horizon scanning exercise. *Journal of Environmental Management* 250, 109479. DOI:10.1016/j.jenvman.2019.109479

Stoate, C., Boatman, N. D., Borralho, R., Rio Carvalho, C., de Snoo, G. & Eden, P. (2001) Ecological impacts of arable intensification in Europe. *Journal of Environmental Management* 63, 337–365.

Stoate, C., Morris, R. M. & Wilson, J. D. (2001a) Cultural ecology of Whitethroat (*Sylvia communis*) habitat management by farmers: field boundary vegetation in lowland England. *Journal of Environmental Management* 62, 329–341.

Stoate, C., Morris, R. M. & Wilson, J. D. (2001b) Cultural ecology of Whitethroat (*Sylvia communis*) habitat management by farmers: trees and shrubs in Senegambia in winter. *Journal of Environmental Management* 62, 343–356.

Stoate, C. (2002) Behavioural ecology of farmers: what does it mean for wildlife? *British Wildlife* 13, 153–159.

Stoate, C. (2008) Multifunctionality in Practice: Research and Application Within a Farm Business. In: Fish, R., Seymour, S. Watkins, C. & Steven, M. (Eds.) *Sustainable Farmland Management: New Transdisciplinary Approaches.* CABI, Wallingford. 161–168.

Stoate, C., Baldi, A., Beja, P., Boatman, N. D., Herzon, I., van Doorn, A., de Snoo, G. R., Rakosy, L. & Ramwell, C. (2009) Ecological impacts of early 21st century agriculture in Europe – a review. *Journal of Environmental Management* 91, 22–46.

Stoate, C. (2011) Biogeography of Agricultural Environments. In: Millington, A. C., Blumler, M. A. & Schickhoff, U. (Eds.) *The Sage Handbook of Biogeography.* Sage, London. 338–356.

UK National Ecosystem Assessment (2011) *The UK National Ecosystem Assessment Technical Report.* UNEP-WCMC, Cambridge.

UNEP (2010) *Strategic Plan for Biodiversity 2011–2020 and the Aichi Targets: Living in Harmony with Nature.* United Nations Environment Programme, Secretariat of the Convention on Biological Diversity, Montreal.

van Delden, H. (2021) SoilCare project, unpublished data.

Walker, L. K., Morris, A. J., Cristinacce, A., Dadam, D., Grice, P. V. & Peach, W. J. (2018) Effects of higher-tier agri-environment scheme on the abundance of priority farmland birds. *Animal Conservation* 21, 199–200. DOI:10.1111/acv.12386.

2 The long view

People have continually developed new means of exploiting the natural resources that the land provides. Wood has provided fuel and building materials for houses, fences and vehicles, and woodland as a whole has provided
food in the form of fruit, nuts and meat, a foraging area for livestock, and an
opportunity for recreational hunting. Wood fuel has been essential for household heating, for cooking and heating water, and for small-scale industrial
processes such as the smelting and working of iron. Wind and water have
been harnessed to provide additional energy for processing food and pump-
ing water. Water itself has been a vital resource, and producing food from the
land, has ultimately occupied the largest proportion of the area. For most of
our history, the power for food production has come from the people themselves and from animals harnessed to work the land. Very often, the management of one resource has been integrated with that of others. The evolving
management and use of land, the exploitation of available natural resources
and the trade-offs between conflicting demands are explored in this chapter.

This chapter draws on information collated for a social learning project
(Stoate, 2010) which combined the knowledge of local residents in the Eye
Brook catchment with published sources and local specialist expertise. Key
local reference material is provided by Ryder (2006) and Jones (2006, 2007).
Information relating specifically to Loddington includes results of archaeological fieldwalking on the farm (Liddle, 2012) following the protocol described by Liddle (1985).

From hunters to farmers

The first people to settle in the area after the last Ice Age were nomadic
hunter-gatherers who had come across the land bridge between Britain and
continental Europe about 12,000 years ago. They were mobile, following and
hunting red deer *Cervus elaphus* and numerous smaller species, and gathering roots, seeds and fruit. There is some evidence to suggest that heavy clay
soils were avoided by these early people who confined their activity mainly
to the higher ground, especially ridges between stream catchments. For example, evidence of human activity about 10,000 years ago has been found on

DOI: 10.4324/9781003160137-2

the ridge immediately to the north of the Allerton Project farm at Lodding-ton, in the form of flint arrowheads, blades and scrapers that are identical to some that have been found in Belgium, Germany and Holland. The flints were dispersed around a fire which may have been used to soften birch resin to glue arrowheads to shafts. Scrapers deposited further from the fire would have been used for paring animal skins.

Further along the ridge, there is not only extensive evidence of later, Mes-olithic, hunter-gatherer activity but also Neolithic and Early Bronze Age flints, and cropmark evidence suggests the presence of a burial mound. Neolithic flints have also been found on the farm at Loddington. The fieldwalking revealed a scatter of flints across all fields surveyed, but with a concentration on elevated light land in 'School Field' (Figure 2.1) compris-ing flakes, cores, scrapers, retouched pieces, a possible axe fragment and an awl. This almost certainly represents a Neolithic occupation site and there-fore the earliest evidence of human occupation of the farm.

Evidence of early Bronze Age occupation of the farm comes from a flint scatter in the southern part of what is now 'Big Park' (a field immediately south of the current village), including a planar convex flint knife on the ridge top, indicating the possible presence of a burial mound. The ridge to the northeast of Loddington is the site of what may be a Late Bronze Age de-fended earthwork and double ditch field boundary systems dating back about 3,000 years. Pollen records suggest that the landscape remained predomi-nantly wooded through the late Neolithic and Bronze Age periods (3,000–5,000 years ago), with open areas for grazing and small-scale cultivation.

The pollen record from around 3,000 years ago reveals an increase in grassland and cereal cultivation. The development of bronze tools, and later, iron ones, no doubt made the task of clearing woodland easier, and the rate of woodland clearance appears to have increased considerably, includ-ing formerly avoided soils such as boulder clay. Pastoralism predominated, with cattle and sheep being the main species, while barley *Hordeum vul-gare*, emmer *Triticum turgidum* and bread wheat *Triticum aestivum* were also grown. The Iron Age landscape was therefore characterised by a substantial increase in open grassland with some cultivation close to farmsteads. Cattle would have been kept for meat, but also for traction, providing an important source of energy that enabled larger areas to be managed than was previ-ously the case. Trading outside the immediate area was well established by this time, with continental pottery being found in the nearby city of Leices-ter. Other local evidence shows that there were contacts with many other parts of Britain and the Roman Empire.

Roman and Anglo-Saxon periods

There is limited evidence of occupation during the Iron Age, but the pres-ence of pottery sherds clearly indicates that the area continued to be occu-pied. Fieldwalking provides evidence of at least three small Roman sites,

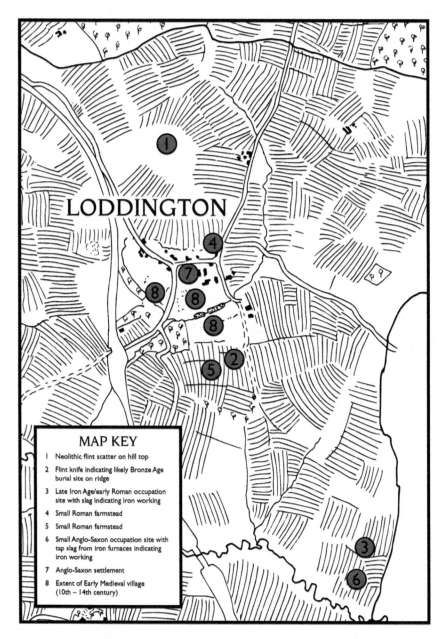

Figure 2.1 Key historical sites at Loddington and distribution of ridge and furrow
evidence of medieval cultivation (drawn from Hartley, 2018)

presumably farmsteads. One of them is close to the site of the Allerton Project visitor centre. Another, in 'Big Park', coincides with aerial photograph evidence of a square shape indicating a ditched enclosure. The third site, in the southeast corner of the farm, includes a scatter of tap slag, as well as pottery, suggesting Roman iron working. Manuring scatters of pottery were also found in the northeast of the farm, indicating Roman agricultural activity.

Woodland was important for pottery and iron smelting and there is some evidence to suggest that kilns (e.g. at Launde immediately to the north of Loddington) and iron workings were sited on woodland edges. One Roman iron smelting site near the village of Belton, neighbouring Loddington but some distance from any known Roman settlement, contained part of a quern from Northern England and fragments of 2nd- to 4th-century pottery, as well as slag spread over an area of about 80 metres.

Evidence from the area to the south, around the Roman town at Medbourne (probably the economic and administrative centre of the area at the time), and other local sites, suggests that several farmsteads and their associated arable land were abandoned during the 3rd and 4th centuries. The more marginal land may have reverted to pasture, or even to woodland.

Anglo-Saxons arrived in England in the 5th century from tribes that were established in a region encompassing southern Denmark and northern France. International contact over long distances is apparent from items such as non-ferrous metal ornaments and even ivory and cowrie shells that would have come from the Middle or Far East, or Africa. Current local evidence suggests a dispersed settlement pattern, probably with a relatively low population during the Anglo-Saxon period. At Loddington, the evidence comprises pottery close to the centre of the modern village. In the southeast of the farm, the fieldwalking revealed Anglo-Saxon pottery combined with a large amount of iron slag from iron furnaces, suggesting an area of Saxon ironworking close to the site of the earlier Roman farmstead.

The medieval open-field system

Major changes took place around the 8th century with the increased adoption of Christianity and a complete overhaul of rural community structure and land use. Isolated farmsteads and small settlements were abandoned in favour of nucleated settlements in the form of villages. This is when the open-field system was introduced in which the land associated with villages was divided into large fields, each of which was further divided into a number of 'furlongs'. These, in turn, were divided into a series of strips, each of which was allocated to a villager. These strips remain clearly visible in many pasture fields today in the form of 'ridge and furrow'. This farming system, and the community social structure associated with it, was to persist for several hundred years.

The strip system encouraged the sharing of oxen between villagers who would plough each strip in turn. Because of this communal open-field system of cultivation, there was therefore a need for synchronisation of cropping.

Each village normally had three large fields and a three-course crop rotation comprising two years in crops and one in fallow (where no crop was grown). There are no local records of specific crops for this period, but a survey of the neighbouring parish of Belton (Cullingworth, 1786), although very much later, revealed a rotation for an open-field system comprising wheat, beans *Vicia faba* and then fallow, and this may have been typical for the area for some considerable time.

The 1786 survey of Belton suggests that there was also sometimes grass within the arable fields in the form of strips of short-term leys and headlands sown with grass which was cut for hay two years in three. In the third year, this grass would have been part of the fallow stage of the rotation which was grazed by livestock after harvest by common right of the members of the community, whether they were cottagers or the lord of the manor. The consequences of this communal grazing were that everyone's arable strips benefited from the dung from grazing animals.

The Domesday Book of 1086 provides valuable information on the structure of rural communities, comprising lords, sokemen and tenants, cottagers and slaves, of which the latter were within the control of the lord. Tenants were required to carry out work on the lord's land, and held only lifetime rights to their land, whereas sokemen were relatively free. As well as rent payable to the lord, villagers were expected to pay tithes to the church in the form of corn, hay, wood, lamb, wool and other products of the land equivalent to 10% of annual production.

The Domesday Book recorded the taxable potential of each township, providing evidence of the population and the resources available to it, including common land, meadow, woodland, wind and water mills, and plough teams. Land area was measured in carucates, one carucate being regarded as the area that could be ploughed in one season by a team of eight oxen (approximately 50 ha). At Loddington, there were 12 carucates of land and 20 acres (8 ha) of meadow, plus an area of woodland. The population comprised nine freemen, seven villagers and seven smallholders. If each of these represents a household, then the population at this time was slightly higher than it is today. There were fourteen and a half plough teams, one of them belonging to the lord.

Much woodland clearance in the local area took place between 950 and 1350, with open fields managed for food production replacing forest and 'waste'. Despite this, woodland was a valuable resource and continued to perform an important function for local people, providing a number of essential resources for fuel, building and forage. Woodland was also the preserve of royalty and the nobility for hunting, including the Ridlington deer park to the east of Loddington, and this created tension between local people and the king. Clearance of woodland, trespass and poaching were carried out despite often severe penalties.

Evidence of medieval cultivation is provided by records of former ridge and furrow, and 1940s aerial photos of the distribution of ridge and furrow can be used to map the minimum area of land under cultivation. This reveals that most of the land area was in cultivation as, although the population of the area as a whole was much lower than it is today, so also were crop yields, so a large area of cropped land was needed to feed the population. Where ridge and furrow have been destroyed by subsequent cultivation, widespread scatters of Saxon, Norman or medieval pottery fragments provide evidence of medieval cultivation. This is because broken pottery was disposed of in manure heaps that were subsequently spread on arable land.

For example, pottery scatters around Loddington reveal that these fields were used for arable cropping from the 10th century to at least the 16th century. The earliest medieval sherds were found closer to the village and later ones further away, suggesting either that the land further from the village was not initially used for arable production, or that manuring did not occur there. A very dense scatter south of Loddington village coincides with earthworks visible from 1940s aerial photographs and suggests that the village was much more extensive in the 10th to 14th centuries than it subsequently became. The village is recorded in the 1130s as being the centre of a small 'hundred' of 48 carucates including other neighbouring villages, but from about this time, it became part of the new priory at Launde to the north. Poll Tax records of 1377 suggest a village comprising 50–60 households, around three times the size of the present settlement.

The mid-13th century to mid-14th century had been a period of relative prosperity. With crops and methods that were, by today's standards, very low yielding, there was a need for a large land area to provide for the population. Pressure on the land was high. In 1348, the Black Death arrived from continental Europe, marking the start of several waves of plague that considerably reduced the population. Despite this, the pressure on woodland for fuel and building of houses and ships continued through subsequent centuries, although the pressure for land clearance for farming was greatly reduced.

Although the reduced size of the population reduced pressure on the land for arable crops, the demand for wool for export remained strong. Some local villages were reduced in size or were totally abandoned in the medieval period. The substantial contraction in the extent of the settlement at Loddington is attributable to the depopulation associated with the plague. In the late medieval period (15th and 16th centuries), the village appears to have been reorganised, with no continuing evidence of occupation to the south of the present village, but some new activity to the north. In contrast, population census data and estimates of population size for Tilton and Halstead, the parish neighbouring Loddington to the west, reveal no evidence of the effects of plague as the population seems to have been relatively stable for most of the millennium.

Enclosure of farmland

From the 12th century, wool had been exported to Italy for the textile trade there, but this export expanded through the medieval period and became an extremely important source of income, both for farmers and for the government through taxes. Taxes on wool increased from the late 14th century and there was a switch to producing wool for domestic textile production by the end of the 16th century. The change from a focus on arable crops to livestock production for wool and meat also stimulated one of the greatest changes to take place in the history of our farmed landscape. The open-field system was abandoned in favour of privately owned enclosed parcels of land. This occurred at Loddington between 1628 and 1630.

A smaller population reduced the demand for arable crops, and therefore arable land and consequently increased the value of pasture for livestock, encouraging the consolidation of parcels of land through the enclosure process. A move away from common fields enabled farmers to breed livestock more selectively and livestock breeds improved throughout this period. The second half of the 18th century also saw considerable improvements in the control of crop rotations, including the adoption of clover *Trifolium* spp. leys and turnips *Brassica rapa* on former fallow land. For those who could afford it, enclosure therefore resulted in an improvement in the land and the farming systems adopted. Enclosure was an expensive process because of the legal fees that needed to be met and because it required the creation of ditches, the planting of hedges, and the erection of fences to protect those hedges. There was a substantial demand for fencing materials from nearby woodland, as well as for young hedge plants.

However, the prospects for those who could not afford to enclose land deteriorated. This was sometimes explicitly acknowledged when the land was set aside for the poor at the time of enclosure. In the neighbouring parish of Belton, the 'Poor's Land' comprised 14 ha and was designated in 1631 when the royal hunting forest of Leighfield was enclosed for farming.

The first half of the 18th century was a period of depression for farming in this area with further loss of small farms and amalgamation of holdings into larger ones. The subsequent fortunes of farming for well over a century fluctuated considerably between periods of growth, for example during the Napoleonic Wars and First World War, and periods of depression such as the 1830s and 1840s. During periods of growth, farmers responded by making further improvements to their land. Records for College Farm which neighbours Loddington to the east show that 5,000 drainage pipes were bought in 1893, a further 1,000 in 1897, and 1,500 more in 1899. The 1881 population census for the neighbouring village of Belton lists a 'land drainer' as being resident at that time, so the period appears to have been one in which the water-logged nature of the heavy clay soils was being addressed.

Glebe land owned by the church provided direct income to the church, while tithes represented a tax on agricultural production that had been levied since the 8th century. Tithes were divided into great tithes and small tithes. Great tithes comprised primarily grain, hay and wood and went to the church institution, whereas the small tithes comprised items such as wool and the annual increase in farm stock and were paid to the vicar. In the 1840s, these tithes were changed so that an annually revised payment was made to the church, based on the productivity of the land. The maps and records that were drawn up to accomplish this provide a valuable record of the landscape and farming in the mid-19th century. It is clear from these that by the 19th century, there was very little arable land, marking a sharp contrast with land use during the medieval period. This low proportion of arable land is also in contrast with the situation today. The 333 ha area occupied by the Allerton Project's farm at Loddington included just 9 ha of arable land when the Tithe map for Loddington was drawn up in 1847. This represents the lightest land on the farm and electrical conductivity mapping of the main arable field, carried out in 2013 when this technology was first being applied, highlights how well the 19th-century farmer had identified the most suitable land for cultivation (Figure 2.2). It is worth noting that this small part of the farm is also the one referred to earlier, where the greatest density of Neolithic flints was found, indicating that this site, with its relatively well-drained, easily worked soil, was the first to be occupied by people.

Large estates became established during the 18th and 19th centuries as small farms were amalgamated into larger units. The estates and large houses were often owned by families whose wealth was not made locally, but from industry, railways, and even sugar cane plantations. Hunting of foxes *Vulpes vulpes* was an extremely important activity in the 18th and 19th centuries, motivating the establishment of fox coverts and other woods in the area, creating local employment for servants, innkeepers and grooms, as well as providing recreation for the wealthy. Nineteenth-century parish records sometimes refer to gamekeepers, providing evidence of formal shoots.

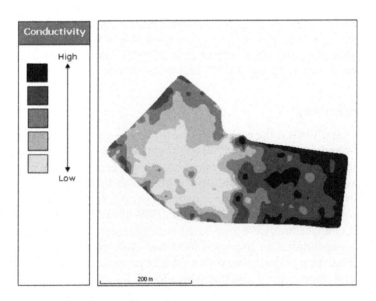

Figure 2.2 Electrical conductivity map of School Field, showing light soils (yellow) in the upper, western end and clay (black) in the lower eastern end

Such shoots are likely to have been on a small scale compared to those further east but were clearly an established component of community culture. They would have depended on a combination of naturally occurring pheasants *Phasianus colchicus* and grey partridges *Perdix perdix* and additional birds reared under bantams for release into woods and fields.

Coal production in Nottinghamshire for domestic use across the region increased dramatically during the 19th century, replacing locally sourced wood, and this was a key driver for development of the railway system, itself fuelled by coal. The railway through Loddington, linking Halstead to the north to East Norton to the south, was established in the 1870s as part of the national expansion of the rail network and served an important role for the transportation of goods such as grain, livestock and milk from local farms, as well as coal, and ironstone which was quarried locally.

Along with well-managed osier beds for basket making, woods across the area continued to be a valuable resource through the 19th and early 20th centuries. In the 19th century, the price of coppice wood and timber increased considerably, as did the price of bark for tanning. Oak was in short supply for shipbuilding and cordwood was in demand as pit props for the rapidly expanding coal mining industry to the north.

The First World War marked a major change within rural communities such as that around Loddington. Many of the woods in the catchment were felled for timber and were subsequently replanted in the following decades. In some cases, this meant clearing the scrub that had been established in the intervening years. As was the case with other large estates across the country, the Keythorpe Estate to the south of Loddington was broken up after the First World War and sold off, often to tenant farmers. These farmers downsized from their rented farms in order to buy their own smaller farms, free of rent. The decades following the First World War proved to be the lead up to a period of even greater change in the management and use of natural resources, following the Second World War.

1930s agriculture

The 1930s and 1940s is the earliest period for which we have been able to obtain first-hand experience of life through the memories of people who grew up during this period. Evidence of farming activity is drawn from interviews with these local residents, providing a first-hand account of life at that time. The 1930s and 1940s represent an important period. In part, this is because it coincides with the Second World War (1939–1945) and with all the major social changes that were associated with that. The war highlighted the country's dependence on imported food and other commodities and food security became a major issue, with a rapidly introduced national policy of agricultural improvement and intensification.

The 1930s also saw the start of a period of increasing reliance on fossil fuels. While coal from the Nottingham coalfields had been used locally for some time before and fuelled the introduction of a railway through the catchment in the 1870s, the 1930s saw the additional and widespread adoption of oil, and electricity generated from coal. This had a large impact both on local people's personal lives and on the way land was managed.

Developments in agricultural productivity accelerated following the Second World War in response to a national need for food security and to the development more widely of new technologies. Whereas fossil fuels had influenced local land use in the 19th century only through the use of coal for railways, the introduction of tractors in the mid-20th century marked a major shift in the way the land was managed.

Interviews with local residents who were children in the 1930s provide an insight into these changes in land use during a period in which fossil fuel use began to dominate farming inputs and operations. They witnessed the change from horse-drawn implements fuelled by oats *Avena sativa* and hay grown on-farm, to petrol- and diesel-powered tractors which could cover ever-increasing areas of ground as the technology developed. Rotations and cultivations for weed, pest and disease control were replaced by chemical applications, synthetic fertilisers became widely adopted, and the area of arable cropping increased considerably. A study of sedimentation rates in Eyebrook Reservoir towards the base of the Eye Brook catchment, downstream from Loddington, indicates that sedimentation rates increased three-fold in the second half of the 20th century as a much larger area of land was cultivated (Foster et al., 2008). This issue is explored further in Chapter 6.

Population census data for the neighbouring parish of Tilton and Halstead reveal that 50% of the increase over the past millennium occurred in the 20th century (Figure 2.3). This more recent increase reflects increases seen across the country and can be attributed to exploitation of resources from the Empire, increased industrialisation, and improvements in agricultural production, hygiene and medicine. Many of these developments were made possible by the discovery and increasingly widespread adoption of fossil fuels for transport, industry and household use, as well as for farming.

Mains water and sewerage were introduced, electricity generated by coal-fired power stations brought power to individual homes, and the occupants of those homes gradually became less connected with the land that surrounded them. Whereas a century ago, most of the rural community had some involvement in management of the land, it is currently less than 2%.

Grass and livestock farming

In the 1930s and 1940s, sheep and cattle farms were considerably smaller than they are today and the breeds, especially for cattle, were generally less productive. The labour input was high, especially for hay making and milking, until labour-saving devices such as milking machines were introduced. Weed control was carried out using hand implements, and grassland was largely unimproved until fertilisers and the ryegrass *Lolium* spp. hybrids

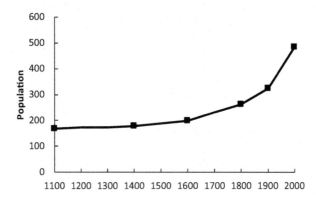

Figure 2.3 Historical population change for the parish of Tilton on the Hill and Halstead (from Stoate, 2010)

Figure 2.4 Hay making in the neighbouring parish of Leighfield in the 1930s. Photo courtesy of Rosemary Richards

that could benefit from them were introduced. The number of plant species present in grassland was higher than it is today, supporting more wildlife.

Hay production (Figure 2.4) was central to winter feeding of livestock and there was no silage and little imported feed. Milk and animals for slaughter were traded locally, and animals that required further fattening were sold through local markets to farmers from further afield. Wool was taken to the local town of Stamford to be graded and used for carpets or clothing.

Sheep and cattle were usually traded locally, either in the town markets of Leicester and Melton Mowbray, or in the market that was occasionally held in the neighbouring village of Tilton which was also an important social event. In either case, farmers would walk their animals to market, although the trains were often then used to transport them to abattoirs or to other farms for further fattening.

BOX 2.1 QUOTES FROM LOCAL RESIDENTS: GRASS AND LIVESTOCK

It was all anthills, that hill. Oh, it was covered in them. You couldn't mow it with a machine. 100 acre around Tiptoe, you couldn't take a machine on there.

JW

You had to go round with a scythe and scythe down the thistles, and with this hassock hoe you would chop off the jack thistles. That grass with cowslips and buttercups and all that sort of thing – it wasn't nourishing for the sheep.

KF

I used to fill my time, when Dad had gone out, with cutting this-tles down in the summer, with a scythe. I used to go up the side of Tiptoe and look where there was a patch of biggish ones, because I used to go round in a circle and cut till you got to the middle and then set out again.

FD

[Dad had] a mowing machine behind the horse. He would be busy sharpening up the knives on the mower. The next thing was the hay rake. You didn't have to fall off, because if you did, and the horse went on, you got caught up in the rakes. The next thing was the swath turner, and that turned the hay. Then eventually, you got it back in rows again. Then you made it into what we called 'cocks' which were just little heaps, and then when [the hay] got drier, you drew these together and made it into cobs. Then you put a big chain round them, and harnessed the horse, and drew the cobs right to where the stack yard was. We children used to ride on top of the cobs.

RM

In that long meadow some nights when we were cobbing there'd be 14 folks in the field.

NF

You stacked the hay then for the winter and it got all solid. Then you had a cutting knife. It was a big thing. You used to sharpen it up, and it was hard work, and you'd cut pieces out of [the hay] and fodder it to the beasts.

RM

These store lambs would go up into Lincolnshire where they would fold them on roots. They used to grow turnips and that sort of thing. They used to make these willow hurdles and fold them

(Continued)

in a certain area in the days before electric fences. There was a firm down in Billesdon that made baskets and that sort of thing but they also made these hurdles and sold them into Lincolnshire. That's where the osier beds came into being, down by the brook.

KF

We had Lincoln Red cattle. Mrs Parker at Rodhill Farm had Ruby Reds, Devon cattle. Mr Button had North Devons. He used to milk these cows and their milk was very rich with cream. [They] wouldn't have an awful lot, about 6 or 7 litres I suppose, and Mrs Button would make the cream.

KF

They was all Lincoln Reds then. They was a good type, but of course, like everything else, they went out of fashion. We had 35. One bull, four or five cows and others were sucklers with calves. When they come big enough we took them off and put some more on. That's how they used to do it in them days.

TW

You'd get the odd Shorthorn creeping in because people tended to move from the Lincoln Reds to something that was giving a drop more milk. You used to have to take your milk down to the station by horse and trap and they'd be all gathered down there at Lowesby station. When the milk train came you unloaded your churns into these railway wagons and bring your empties back. The milk went to London.

KF

Our Friesians originated from the Dutch. Hugh went out to Friesland in 1949. Then he bought cows from well-known breeders in this country. They used to come by train to East Norton, and we had to go to the station to get them.

VK

[Cattle] feed was rolled barley and grass, and a bit of hay. Then the feed changed altogether, with a lot more bought in stuff and that sort of thing.

TW

The fattening cattle always had linseed cake. We'd roll a certain amount of oats and mix the dairy cake with it. Sometimes when we'd harvested oats and beans it was always a job to get them dry and we'd feed the sheaves to them, taken out of the stack.

KF

There'd maybe be four or five [people] milking. There'd be a 10 and a 12 and a 14 in the shed, hand milked before we had a milking machine. That was revolutionary. You only wanted one person to milk. All of it had to be carried over the yard to be put over the cooler.

NF

Tilton sales were in the spring and the autumn mainly, and then they'd have one, say, every fortnight. Say, three sales in the spring and three in the autumn. I often walked up with the sheep. Parkers had a man who would go round to the farms and he'd buy them and then he'd send his drovers to drove them.

FD

Mostly it was more cattle in the autumn, coming in off the grass. There was also a big sheep sale in the autumn when you hadn't got the fodder for them and it was better to sell them in the market, straight out of the fields. The cattle were ready for slaughter so it would have been dealers and butchers that used to buy them and they'd be moved on straight to the abattoir.

KF

For Melton market we used to take sheep and put them in a field at the top of the hill. Charlie used to take quite a few sheep in and drop them in this paddock and he'd leave a note on the gate and their drovers would come up in the morning and walk them into Melton.

FD

You used to drive cattle into Melton and Leicester and they used to be all down the town, in bunches, right down the street. They kept moving them up as they sold them. Then they used to load them on the railway.

JW

We used to take ours to Leicester and we used to walk them down to Thurnby, and then the next morning we'd get up about five and walk them from Thurnby to Leicester Cattle Market. The auctioneers always had fields around the towns you see, so you could move your cattle there. It wasn't a hard job because there were no traffic. You would walk at about four mile an hour. It wasn't long getting there.

JW

Arable farming

Most of the land was pasture in the 1930s, with a few crops grown to provide feed for livestock. During the Second World War, the County War Agricultural Committee (popularly known as the 'War Ag') intervened to ensure pasture was ploughed up for crop production (Figure 2.5), either by the farmers themselves, or if those farmers were not equipped or inclined to do it, by others appointed to do so. Soil fertility depended on the spreading of manure produced on the farm until the widespread use of synthetic nitrogen fertiliser (derived from fossil fuels using the Haber-Bosch process) during this period. Weed control was by mechanical means (repeated hoeing) until the arrival of the first herbicides (Stoate & Wilson, 2020). Tractors were rapidly adopted during and immediately after the Second World War. The arrival of tractors marked the end of a long dependence on horses that were fuelled by locally produced grass and oats, reducing labour requirements. An enormous amount of labour was still required, especially at harvest for cutting, carting and stacking, and later in the year, for threshing.

Figure 2.5 Ploughing in the local parish of Great Easton in the early 1940s. Photo courtesy of Phil Johnson

---

BOX 2.2 QUOTES FROM LOCAL RESIDENTS: ARABLE FARMING

I couldn't see a field that was ploughed except Wrights at Oxey. They had a patch down the side of the road which they used to grow kale in for the cows.

FD

Coldborough was the first field Holmes ploughed up and that was ploughed with steam engines. [I] used to stand and watch them from the top of the lane there. [They] ploughed it crossways. You couldn't buy tractors in those days. Very, very few about. The War Ag got most of them and you had to go to them to get your work done.

PW

There was always a bit of muck spreading to do. When we cleaned the cows out we used to put it in a heap, outside the door for a start and then we'd come up with the horse and cart and fill the cart and tip it down the field. Then you'd got to spread it.

FD

We used to grow ten or twelve acres of roots, swedes, cabbage for the cows. You had a job to pick them up, they were that big. We used to muck it very heavy. We used to grow kale but we grew more cabbage because it was easier to cart out to the farms. We grew a lot of oats. Potatoes. We were ordered what to do in war time you see. You had Ministry men come round and say we want this in this field and this in this field.

JW

There was only one arable field at that time. It was 28 acres. It was actually four fields in a rotation. There'd be oats and beans that would be ground up for the [live]stock and that sort of thing, wheat for the straw, oats for the cattle. Cabbages were for the winter feed. Then there were mangels.

NF

The weeds got too bad on the third year [and] you cultivated it as a fallow. You used to go out there and scuffle it. At the end of the season when it was ready for sowing it would be a nice fine loam.

KF

[For fallow] we used to rip [the ground] up mostly May time, sometimes earlier than that. Then we used to do nothing else but plough it, harrow it, whichever it was. Then we used to drill that with wheat. That used to go in early. Then we used to always have a good crop of wheat. In those days there was no artificial fertiliser.

RG

You had the horse hoe to keep the weeds down. You wouldn't do it many times. There wasn't the weeds about as there is now. We used to binder it and bring the stacks down here [to the farm]. In those days, when you'd finished threshing at the end of the day my father used to get all the [weed] seed and take it out and burn it.

NF

(Continued)

My dad used to mow all around the field with a scythe. He wouldn't drive the tractor through the corn. Then you used to go in with the binder and there used to be one sitting on the tractor and one sitting on the binder. We used to do about ten or twelve acres in an afternoon you know. Cut it and stook it all up. Stooking wasn't a very nice job. You'd pick two sheaves up and go and stand them up and then go and pick two more because they were all over the place. We'd have six or seven [people working]. We had to go along and pitch them onto the trailer, and then take them and make a stack, and then you used to thatch it.

JW

We had a fortnight off at hay time, then back to school, and then another fortnight off at harvest time carting the horse and cart in them days – stacking the corn up into ricks like, into corn stacks.

PW

There were a lot of people who were good stackers and they used to thatch as well you see. You used to have a bundler behind the thrashing machine and he'd chuck these big bundles of straw out and then you'd chuck it in a heap and then you had to draw the straw together to get the thatch out of it. Then you had some pegs, hazel pegs.

NF

[Threshing] were a dusty job. They always got to us in about June time when all the dust was in it. The threshing drum was down [in] Norfolk. They'd do the farms down there first.

JI

When we got onto thrashing and that sort of thing we'd go and help one another. We used to go down a lot to Loddington. We had perhaps 10 or 12 stacks in the [Big] Park Field. We used to go stooking round the park and it'd take you all morning to go round the outside. There'd be eight or so of us.

NF

BOX 2.3 QUOTES FROM LOCAL RESIDENTS: HORSE POWER

You had a heavy horse that took the big cart, when you were carting manure and that, and then had a vanner that took the float, and you usually had one you could ride. We used to get in the float and go up with the horse and we'd play all morning in the fields while Dad worked.

RM

We always used to go round in a pony and a float, round the fields, feeding cattle and that. We used to winter them outside and feed them under the hedge. Two of us used to do it. You'd keep a good cob sort of horse for that purpose. There used to be a haystack in each field in them days – all loose hay then. We had four or five horses of one sort or another.

TW

When you'd ploughed a field, you'd walked some acres you know. It wasn't [so bad] when you was doing the grass because you sat on the mower. When he was ploughing he used to have to walk behind.

EP

The land was that heavy that only one furrow was hard graft for a pair of horses, and it was hard graft for the man as well to hold the plough in an upright position. To plough an acre a day was probably the biggest amount they could plough. That field was about 25 acres so he'd be working there very close on a month.

KF

Once we got them tractors, we used the tractors for everything. The horses went.

JW

Tractors, they made it really, didn't they? There was a lot more work done on the farm with the tractors than with the old horses.

TW

Woodland

Woods were important for hunting and shooting, and were valued by children as a playground. The shooting of pheasants and other game does not seem to have been a major activity locally, even immediately before the war, but organised shoots were held in the parishes of Allexton and Keythorpe, to the south of Loddington, with pheasants being reared under bantams and released in the early autumn. Fox hunting, on the other hand, continued to be an activity that had far-reaching influences on the landscape through the management of hedges and planting and management of 'coverts', and on rural life.

Woods were also a source of timber. Oak *Quercus* sp. was cut for fence posts, and riven rails were made from ash *Fraxinus excelsior*. Hazel *Corylus avellana* was cut for thatching pegs and stakes and binders for hedge laying. Other sticks were used to fuel stoves and for pea sticks and bean poles, and logs were used for open fires. Even the main mode of transport relied on timber from the woods, with planks for making and mending carts and timber for wheel-righting coming from local woodland.

BOX 2.4 QUOTES FROM LOCAL RESIDENTS: WOODLAND

I used to help with the shoot. When Mr Hoare was alive we used to have three or four shoots a year. That's all we used to have. We used to go in with Park Farm, and they used to join in with Allexton. Park Farm always used to be a Boxing Day shoot. If we got 20 brace [40 birds] we had had a good day. We used to rear a few the old-fashioned way with the banties and chickens one way or another, and then at Allexton, on our home ground, we used to turn out about 150 there most years.

RG, 1950s

We used to keep the rides clear, cutting them back so that you could drive through. We used to do that for the hunt and for the sake of the woods as you might say, to keep them tidy. Allexton Wood used to belong to the Fernie Hunt but then the Forestry Commission felled it and they took it over.

RG, 1950s

Father said your job will be to cut the poles down in Merrible and Bolt [Woods] and cart them to Uppingham, a load each Saturday, and I used to run behind the old drey, most Saturdays. Me and my sister used to stop at [Beaumont Chase] to drink out of that trough.

JR

Everything he took out of that wood, he took out on an old horse and cart. He kept his horse on the bottom field of Holmes's. It was just a flat cart with four wheels and four corner posts. He used to fetch his old horse in every now and again and load it up and take it out to the road for others to pick up. He made no mess at all in that wood. He could move around there and take everything out – brilliant really.

PW

There was a tin shed and Jack Pepper kept his saw mill in there and an engine to drive it. They used to buy trees and saw them into planks. They were wheel wrights and would repair your shafts in your cart or [put] new bottoms in your wagon. He'd go round and buy an oak here and there, locally. Then he'd store it there to dry out.

KF

I've heard old Jack Holmes say when him and old Walter Wright were young they used to get a bit competitive between them, you know, to see who could make a gate and hang it, all in one day. Take some doing, wouldn't it?

PW

Household food, water and sanitation

Food was largely produced in the home, and most of the rest was produced locally. The household pig was often central to meat consumption and made use of waste from the kitchen. Most villages had a butcher, or at least someone who would make the annual visit to kill and butcher the household pig, and the nearest abattoir for cattle and sheep was just seven miles (11 km) away. Milk from dairy or household cows was used to make butter, especially if there was a surplus. The seasonality of food production determined what was eaten when, although some was stored for use later in the year. Pork was salted, and most people grew their own fruit (which was often bottled for the winter) and vegetables, and many kept poultry. Most villages had their own bakery.

Household water was from wells and pumps which, for some, were located next to the house, but for most were a short walk away. Water had to be collected each day, or if baths were required or washing was to be done, several times during the day. In some cases, hydraulic rams used the power of the water flow from a stream or spring to pump the water up to the houses or to livestock drinking troughs. Rainwater was also collected from roofs and this 'soft' water was preferred for washing clothes.

Sewage treatment works were built for local villages in the mid-20th century. For houses not on a mains sewer, such as those in the hamlet of Loddington, septic tanks were gradually adopted from around the same time so that flush toilets became much more common. This improved hygiene but increased the flow of wastewater from houses to the stream, as discussed in Chapter 6.

BOX 2.5 QUOTES FROM LOCAL RESIDENTS: HOUSEHOLD FOOD, WATER AND SANITATION

Baker Bird used to deliver on horse and cart. He made [the bread] in the morning and delivered it after dinner and it was still red hot. … His bread was lovely.

EP

You used to have to turn the churn round and round and round, and sometimes it was just as if [the butter] wouldn't come because you had to have it at such a temperature. The colder it was, the better the butter came.

RM

Mother used to pickle eggs. She used to use something called water glass. She had a big pippin – it used to be full of eggs. When you came to get them out, this water glass had gone sort of white and scrunchy. The eggs went for cakes and puddings.

RM

The shelf in the [pantry] was full of bottled fruit. She bottled plums and damsons. You didn't bother with apples – they used to keep ever so well down in the cellar.

FD

He'd come and kill [the pig] one day, come and cut it up the next day … [It was] salted for three weeks. A week and a half you put it in the salt, then you'd turn it over and salt the other side. Then you'd fetch it out and clean it all down, and hang it up for bacon.

PW

We had a pump - spring water – in the farmyard across the cobbled yard. We had to carry two buckets of water [each day] … across to the house and empty them into pancheons. They'd hold about a bucket and a half each and they stood under the shelf in the dairy so we got plenty of water all day if you fetched it in the morning. Two or three buckets a day. Only boiling kettles, you didn't use a terrific lot of water.

JG

We used to get the water in for washing on a Monday from the soft water tank in the scullery place. It was water off the [roof] slates.

NF

We used to carry the water for these cottages out of the well up to a wash house at the top, with a copper which was shared by two other neighbours. No taps anywhere.

VF

We had the toilet outside in the shed. You had to come out of the door, around the path, in the middle of the night. It was a pan [and] you had to dig a hole and bury it every week. We put it in the field.

FD

There were lots of cottages with pans. There was a cart that came round once a week and they emptied these pans into it.

KF

Historical changes in wildlife abundance

The abundance of wildlife has changed in response to these historical changes in landscape and its management. The Leicester Museum curator and taxidermist, Montague Brown provides a record of vertebrates present in the 19th century (Brown, 1889). Pine martins *Martes martes*, polecats *Mustela putorius*, Red kites *Milvus milvus*, ravens *Corvus corax* and buzzards *Buteo buteo* all occurred locally at the start of the century but were no longer present by the middle of it. At the end of the century, corncrakes *Crex crex* were described as still being 'generally distributed', while red-backed shrikes *Lanius collurio*, redstarts *Phoenicurus phoenicurus* and corn buntings *Emberiza calandra* were 'sparingly distributed'. None of these species occurred locally by the time the Allerton Project started a century later. Nightingale *Luscinia megarhynchos* and grasshopper warbler *Locustella naevia* were previously 'sparingly distributed' in local woods and thick hedges but now very rarely occur locally. Grey partridge was valued as a game species and described by Brown (1889) as being 'resident and common'.

Towards the end of the 19th century, willow warblers *Phylloscopus trochilus* and chiffchaffs *P. collybita* were both described as 'commonly distributed'. Brown (1889) considered blackcap *Sylvia atricapilla* to be 'sparingly distributed', while garden warblers *Sylvia borin* were 'generally distributed, breeding and more common than the blackcap'. The relative abundance of the two species is very different today, with blackcap having increased by 335% nationally between 1970 and 2018 (Defra, 2019), whereas garden warbler numbers declined nationally by 11% over the same period. From approximately equivalent numbers at Loddington in 1992, blackcaps and garden warblers are now present at a ratio of 20 to 1.

We are able to quantify these more recent changes as national and regional trends in bird numbers have been well documented through volunteer surveys coordinated by the British Trust for Ornithology since the 1960s, and birds have been intensively monitored at Loddington since 1992. Nationally, species associated with farmland have declined more than those of other habitats, reflecting the impacts of changing land use on the ecology of agricultural land and stimulating much of the research reported in Chapters 4 and 5. For example, between 1970 and 2018, linnet *Carduelis cannabina* and skylark *Alauda arvensis* each declined by 56%, tree sparrow *Passer montanus* declined by 90%, turtle dove *Streptopelia turtur* by 98% and yellowhammer *Emberiza citrinella* by 60% (Defra, 2019). In woodland, spotted flycatcher *Muscicapa striata* declined by 88%, song thrush *Turdus philomelos* by 49% and bullfinch *Pyrrhula pyrrhula* by 38%.

Hickling (1978) described turtle dove as a 'fairly common summer migrant' in Leicestershire in the 1970s, and redstart as having a 'stronghold in our county [in] the Eye Brook valley' with about 90 pairs in eastern Leicestershire in the 1960s. Tree sparrow was described as a common resident. Both local and national data demonstrate a clear decline in numbers of bird species such as turtle dove and tree sparrow since the 1970s. However, towards the end of the 19th century, Brown (1989) regarded these two species as 'sparingly distributed' and the former 'a comparative rarity in the county'. The 1970s benchmark for modern records is therefore an arbitrary one as abundance of species will have fluctuated through history in response to a wide range of influences.

Ravens and buzzards which were very rarely recorded in the 1970s (Hickling, 1978), and red kites which were a scarce vagrant, are now common as breeding species, the latter having been reintroduced. Goldfinches *Carduelis carduelis* were described by Brown (1889) as 'sparingly distributed', but by Hickling (1978) as a 'common resident' and have increased further in abundance nationally since then (Defra, 2019). Other species such as little egret *Egretta garzetta* have colonised the area as part of a northwest range expansion associated with climate change.

The recollections of local residents provide valuable insight into land use and other aspects of rural life in the period leading up to the 'Green Revolution' and complement a range of other sources of information on much earlier periods in history. The historical context of these first-hand reports is clearly understood, both in terms of earlier history, and in terms of subsequent developments. Detailed wildlife records, as exemplified for birds, provide a more objective assessment of changes in species abundance, albeit accepting an arbitrary benchmark.

A purely anecdotal approach to assessing changes in wildlife can be misleading, especially with regard to aspirations for nature recovery, as there is strong evidence that our personal recollections of wildlife abundance in our youth influence the scale of our expected reversal in population declines. The expected abundance does not exceed that of our own experience, even though it may have been much higher in the past. Known as 'Shifting Baseline Syndrome' (Jones et al., 2020), this can have cumulative impacts on our individual and collective expectations of wider environmental performance across generations.

Establishing aspirations based on earlier periods in history as a benchmark can be even more difficult. As referred to previously, pollen records provide an insight into plant community composition thousands of years ago, including the species composition of woodland and relative abundance of grasses. Sediment cores taken from Eyebrook Reservoir, towards the base of the Eye Brook catchment reveal a three-fold increase in the rate of sedimentation in the second half of the 20th century, reflecting increased soil erosion as the arable area increased (Foster, 2008). The same samples also reveal a decline in the spores of field mushrooms, a species associated with pasture, especially that grazed by horses.

King (2008) used longer-term historical land-use maps as a focus for discussion with water quality experts to estimate the extent to which water quality, and by implication, aquatic ecology in the Eye Brook had changed over the past 800 years. For this, she used 1940s aerial photos of ridge and furrow to map the minimum area of land under cultivation in the medieval period, and 1840s Tithe records to map land use for that period. There was a consensus that, although the land area was largely arable in the medieval period and pasture in the 19th century, with a large increase in the arable area in the second half of the 20th century, there is unlikely to have been a direct relationship between the arable area and water quality and stream ecology.

For many centuries, livestock would have had unhindered access to streams, and villages close to streams are likely to have deposited human waste into them. However, the low intensity of arable management in the medieval period, including a vegetated fallow stage in the rotation, would reduce negative impacts of cropping. Similarly, lack of hydrological connectivity would reduce negative impacts of both arable land use and human settlements on water quality. The lack of field drains until the late 19th century, and the lack of sewage treatment works until the mid-20th century partially isolated human activity from the stream. Despite the efforts of today's water companies, the universal adoption of the flush toilet massively increased the discharge of nutrients derived from human waste into watercourses. Undrained land close to watercourses would also have been difficult to manage in the past, creating marshy buffers which further protected the stream and attenuated flow.

On the Water Framework Directive Ecological Status scale ranging from 'Excellent' through 'Good' to 'Moderate' and 'Poor', the Eye Brook was classified as 'Good' in 2008. An understanding of land-use history, combined with our current understanding of hydrological processes, suggests that historical Ecological Status may have been 'better' than what we now regard as 'Excellent'. We cannot know for sure.

We also explored the influence of more recent historical land-use change on wild bee abundance and species richness using transect data from Szczur et al. (2021). Rayner et al. (2021) again used the 19th-century Tithe maps to provide an indication of the proportion of arable and grassland in the 1920s and 1930s, given the relatively low level of local land-use change during this period. Ordinance Survey maps for 1952 were used to provide data for hedge length for the same period as local anecdotal evidence suggests only limited hedge removal in the 1940s. These historical records suggested a 44% decline in grassland area and a 76% decline in hedge length between the 1930s and 2020, while the arable area had increased six-fold. The modelling suggested a 37.0% decline in ground-nesting bumblebees, a 35.7% decline in cavity nesting solitary bees, and a 24.5% decline in ground-nesting solitary bees. This exercise provides evidence for the impacts of landscape change on the abundance of wild bees, but interpretation about the scale of population changes is difficult because of other factors operating over the same period. For wild bees, increased use of pesticides is known to have reduced numbers (Goulson et al., 2015), and changes in abundance of wild bee nest predators in recent decades may also have had an influence that is not captured by land-use modelling (Roberts et al., 2019) so that we can expect the scale of the decline to be greater than that predicted by the modelling. When applied to the tree nesting bumblebee, *Bombus hypnorum*, the model identified a 29.3% decline in abundance but we know that this species was absent from the UK in the 1930s, only colonising the local area in around 2009 in response to climate change.

Climate change, whether long-term over millennia or resulting from anthropogenic influence over the past century, alters the expectations we can have for managing the aquatic environment, or plant and animal communities in the future. Neither farmed landscapes nor the abundance of wildlife within them is timeless and our objectives for them are, in a sense, arbitrary, as we adapt them to the demands that our own species makes on the landscape. Although we may understand what is meant by 'cultural landscape' designations associated with different parts of the country, these tend to be associated with limited discrete relatively recent periods in that landscape's history. If our multiple objectives for landscapes are to accommodate sustaining both us and other species, we need to be aware of the historical context, but also to have a sound understanding of the processes that are currently operating.

References

Brown, M. (1889) *The Vertebrate Animals of Leicestershire*. Midland Educational Company Ltd, Birmingham and Leicester.

Cullingworth, W. (1786) A Survey and Terrier of the Estate of the Right Honourable George Earl of Winchelsea and Nottingham at Belton in the County of Rutland. Leicestershire Record Office DE 585 7/1/79a.

Defra (2019) *Wild Bird Populations in the UK, 1970–2019*. Department for Environment, Food and Rural Affairs, London.

Foster, I., et al. (2008) An Evaluation of the Significance of Land Management Changes on Sedimentation in the Eyebrook Reservoir since 1940. Unpublished report.

Goulson, D., Nicholls, E., Botías, C. & Rotheray, E. (2015) Bee declines driven by combined stress from parasites, pesticides, and lack of flowers. *Science* 347, 1255957. DOI:10.1126/science

Hartley, R. F. (2018) *The Medieval Earthworks of South & South-East Leicestershire.* Liecestershire Fieldworkers, Leicester.

Hickling, R. (1978) *Birds in Leicestershire and Rutland.* Leicestershire and Rutland Ornithological Society, Leicester.

Jones, E. (2006) The last hunters and gatherers of the Uppingham Plateau: some Palaeolithic and Mesolithic sites and findspots in Rutland. *Rutland Record* 27, 243–268. Rutland Local History and Record Society.

Jones, E. (2007) *The Oakham Parish Field Walking Survey – Archaeology on the Ploughland of Rutland.* Published privately by Elaine Jones.

Jones, L. P., Turvey, S. T., Massimino, D. & Papworth, S. K. (2020) Investigating the implications of shifting baseline syndrome on conservation. *People and Nature.* DOI:10.1002/pan3.10140

King, P. H. (2008) *Discuss the concept of "Good Ecological Status" in the Eye Brook Stream, Leicestershire, through an interpretation of historical landuse practices within the Eye Brook Catchment.* Unpublished MSc thesis. University College London.

Liddle, P. (1985) *Community Archaeology: A Fieldworker's Handbook of Organiza-tion and Techniques.* Leicestershire Museums, Leicester. 38pp.

Liddle, P. (2012) *A Fieldwalking Survey of Loddington, Leicestershire. Fourth Interim Report 2009–2012.* Unpublished report.

Rayner, M., Balzter, H., Jones, L., Whelan, M. & Stoate, C. (2021) Effects of improved land-cover mapping on predicted ecosystem service outcomes in a lowland river catchment. *Ecological Indicators* 133, 108463. DOI:10.1016/j.ecolind.2021.108463

Roberts, B. R., Cox, R. & Osborne, J. L. (2019) Quantifying the relative predation pressure on bumblebee nests by the European badger (*Meles meles*) using artifi-cial nests. *Ecology and Evolution* 10 (3) 1613–1622. DOI:10.1002/ece3.6017

Ryder, I. E. (2006) *Common Right and Private Interest: Rutland's Common Fields and Their Enclosure.* Rutland Local History and Record Society, Oakham.

Stoate, C. (2010) *Exploring a Productive Landscape – From a Long History to a Sus-tainable Future in the Eye Brook Catchment.* Game and Wildlife Conservation Trust, Loddington, Leics.

Stoate, C. & Wilson, P. (2020) Historical and Ecological Background to the Arable Habitats of Europe. In: Hurford, C., Wilson, P. & Storkey, J. (Eds.) *The Changing Status of Arable Habitats in Europe.* Springer Nature, Cham. 3–13.

Szczur, J., Rayner, M. & Stoate, C. (2021) *Wild Bee Abundance and Species Richness, Survey Data from Leighfield Forest, Leicestershire and Rutland, 2020.* NERC Environmental Information Data Centre. (Dataset). DOI:10. 5285/4f4d2ef6-2f31-4aa6-aa1a-fe79cc6d6b04

3   Soil

The life support system

For such a small part of the planet, soil plays an incredibly important role in multiple functions, including the production of our food, but also sustaining wildlife, controlling water quality and flood risk, influencing greenhouse gas emissions and sequestering carbon. For this reason, soil degradation has become an increasing global concern, in particular, because of erosion by wind and water and physical degradation through compaction and loss of organic matter.

The susceptibility of soil to different influences varies considerably according to soil texture. Sandy and silty soils are susceptible to erosion by wind as well as water, while clay soils are susceptible to water erosion and to compaction. The clay soils at and around Loddington present challenges because they become wet quickly and dry hard in the summer but their tendency to crack when dry means that they can naturally restructure. Compaction of clay soils, for example by heavy machinery or livestock in wet conditions, limits the rooting capacity of arable crops and grass and their ability to take up nutrients and increases the risk of runoff and erosion, but soil chemistry such as pH and availability of nitrogen, phosphorus and other nutrients also influences crop growth.

Land use and soil life

Soil 'health' and soil's ability to perform natural functions, is largely dependent on biological activity, something that has only been widely acknowledged relatively recently. Soil organic matter, derived from above- and below-ground animal and plant material, is a source of nutrients for living animals and plants, and plays an important part in water infiltration and moisture retention. It is also a major driver of cation exchange capacity, influencing the soil's ability to hold positively charged ions such as potassium and magnesium, soil structural stability, nutrient availability, pH and fertiliser behaviour in soil. Clay soils are associated with relatively high cation exchange capacity, but this is increased further when organic matter is high.

Earthworms and potworms play important ecological roles within the soil, breaking down organic matter so that it is more accessible to

DOI: 10.4324/9781003160137-3

microorganisms, and digesting fungi and microorganisms, thereby making nutrients available to plants. Earthworms also increase soil porosity and aeration, countering the anaerobic conditions, which favour bacteria, and improving water infiltration rates. While slugs and snails are clearly pest species if abundant, some beetle species associated with undisturbed soil are predators of pests such as the eggs and immature stages of slugs, as well as aphids and other insect pests.

On the farm at Loddington, Hamad (2018) highlighted, amongst other things, the important influence of compaction and organic matter on soil biology and therefore on soil function. He chose sites in ploughed fields where there were visible signs of poor crop performance that was most likely due to compaction. He compared them with randomly selected sites in direct-drilled or reduced cultivation fields (where crops had been drilled directly into the stubbles of previous crops with little or no cultivation), low-input permanent pasture, and ancient semi-natural woodland which may never have been agricultural land at any time in the past. The study therefore adopted compacted ploughed land as the most degraded form of land use to compare with the other three categories. It did not attempt to compare representative examples of ploughed and direct-drilled land, but this topic will be discussed in detail later in this chapter.

Clay content in Loddington's pasture and arable soils ranged from 20% to 50%, and from 50% to more than 60% in woodland, with significantly higher clay content deeper in the soil profile in all cases. Penetration resistance data collected using a penetrometer confirmed that compaction was greater at the apparently compacted sites than at the others. Compaction at randomly selected sites in direct-drilled fields was similar to that in woodland to a depth of 30 cm, at which point it increased considerably to levels equivalent to that in compacted plough, probably because of historical ploughing of the fields that are now direct-drilled.

Porosity was significantly lower at the compacted sites than in the direct-drilled fields, pasture or woodland sites. It did not differ significantly between these three land uses but was marginally higher in woodland and most variable in pasture.[1] Soil organic matter declined with depth in the soil profile in all land uses, but was significantly highest in woodland at around 20%, compared with 3–6% in the other land uses. pH was slightly lower in pasture (around 7) than arable (7.5) but was significantly lowest in the ancient semi-natural woodland, at around 4.5.

Thirteen species of earthworm were identified, representing the range of deep burrowing 'anecic' species, shallow burrowing 'endogeic' species, and 'epigeic' species which live in the surface leaf litter layer. Each of these functional groups plays a different role in transporting nutrients and organic matter through the soil profile and making nutrients available to crops.

Earthworm species richness and diversity were significantly lowest in the ancient semi-natural woodland, compared to agricultural land, whether pasture or arable. Twenty-four percent of this variation in earthworm species richness across land uses could be explained by physical properties of the soil such as organic matter, clay content, hydraulic conductivity, and porosity. The low species richness and diversity in woodland were also found for other macro-invertebrates. Within the agricultural land, overall invertebrate species richness and diversity were highest in pasture, intermediate in direct-drilled fields, and lowest at the compacted plough sites.

In terms of abundance, overall invertebrate densities differed significantly between all habitats in the order pasture>direct drilled>plough>woodland. For juvenile earthworms and adults of one earthworm species (*Lumbricus rubellus*) their density and biomass declined with increasing compaction. *L. rubellus* is a very widespread epigeic species that is generally regarded as being tolerant of a range of soil conditions. Even for this species, along with others, physical soil compaction appears to be limiting recruitment and abundance. The relationship with compaction was also found for some other macro-invertebrates. For example, across all sampling sites, the highest densities of potworms, Geophilomorpha (soil centipedes), beetles and Scolopendromorpha (surface centipededs) were associated with the least compacted soil. In line with this, densities of potworms, juvenile earthworms, adult earthworms *Eisenia fetida* and *Aporrectodea caliginosa*, beetles, slugs and snails were all higher in the direct-drilled fields than at the compacted ploughed sites. Of the two earthworms, *E. Fetida* is an epigeic species that is associated with leaf litter and so is likely to benefit from the accumulation of plant material on the soil surface in direct-drilled fields, while *A. calignosa* is a shallow burrowing endogeic species that is susceptible to soil disturbance in plough-based systems. The shallow-burrowing endogeic earthworm, *Allolobophora chlorotica* was present in similar densities in direct drilled and pasture sites.

Overall earthworm biomass was highest in pasture, but biomass of macro-invertebrates excluding earthworms was similar in permanent pasture and direct drilled sites (Figure 3.1).

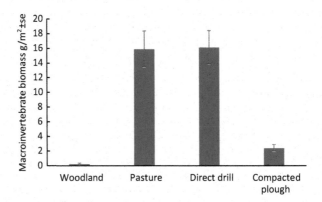

Figure 3.1 Biomass of macro-invertebrates, excluding earthworms, in relation to land use (drawn from data in Hamad, 2018)

The same study also looked at organic matter decomposition rates as a measure of biological activity within the soil. These were significantly higher in direct-drilled fields than the other land uses, and significantly lowest in woodland and at the compacted ploughed sites. Use of mesh excluder nets of a range of sizes revealed that these differences were primarily due to macro-invertebrates, rather than meso- or micro-fauna, although this difference was reduced in the summer when there was an increase in mesofauna activity. Across all the sites, there was a negative relationship between biological activity and compaction. CO_2 flux was measured as an indicator of respiration and was highest in pasture and lowest in woodland, reflecting the findings of the invertebrate surveys (Figure 3.2). Macro-invertebrate biomass explained 15%, and earthworm biomass explained 11% of this spatial variation in soil respiration.

Low levels of biological activity in the ancient semi-natural woodland sites can be explained by the low pH, despite being otherwise suitable in terms of low compaction and high organic matter. Earthworm species vary in their tolerance of pH, but national records show that many species occur at higher densities in woodland than we have recorded where soil pH is higher (Sheppard, 2014). Beyond the influence of pH, this study clearly demonstrates the extent to which compaction limits soil biodiversity and activity, inhibiting soil health and function. The similarity between pasture and direct-drilled fields is interesting in that it indicates that soil biodiversity and biological activity in arable soils can be comparable to those in semi-natural habitats if the appropriate management is adopted.

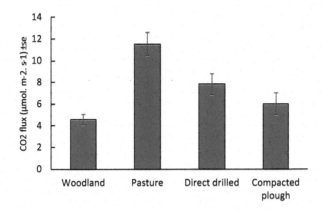

Figure 3.2 Carbon dioxide flux in relation to land use (based on data from Hamad, 2018)

The role of vegetation

Erosion plot research at Loddington has identified vegetation cover as a key influence on runoff and erosion rates (Cooper, 2006). This issue arises primarily as a result of the soil being exposed to rainfall between the harvest of one crop and the establishment of a following spring-sown crop. Cover crops planted to protect soil from erosion between the harvest of one crop and the spring sowing of another one, reduce impacts on watercourses through the transport of sediment and nutrients to water, and have been shown to benefit soil structure and function. The evidence for this last point is less clear on clay soils, especially in relation to agricultural production. As a result, this was investigated at Loddington in two replicated experiments, the first with mixtures of cover crop species (Crotty & Stoate, 2019), and the second with single species (Lee et al., 2017). The mixtures comprised combinations of oats (*Avena sativa*), oil radish (*Raphanus sativus*), vetch (*Vicia sativa*), and *Phacelia* (*Phacelia tanacetifolia*), while the species adopted for the single species experiment also included buckwheat (*Fagopyrum esculentum*). In both experiments, plots with cover crops were compared with bare stubble control plots.

Soil magnesium was lower, and nitrate was higher in the most complex mixtures (oats, *Phacelia*, vetch and radish) than the control bare stubble plots. Epigeic earthworm abundance and biomass were also higher in this mixture, and in oats, *Phacelia* and radish, than in the control plots and oats and *Phacelia* alone. This was accounted for mainly by differences in *L. rubellus* and *A. caliginosa*. Of the arthropods, millipedes were more abundant in oats, *Phacelia* and radish than the other plots, and flies were more abundant in oats and *Phacelia* than the others. Importantly from a farming perspective, slug and snail abundance did not differ between cover crop mixtures, or between these and the bare stubble control plots.

There was a tendency for grass weed biomass to be associated with the lowest cover crop biomass – vigorous cover crop growth suppressed weed growth. This relationship followed through to weed biomass in the following spring-sown oats crop, with the highest grass weed biomass in the plots that had previously been bare stubble controls, and lowest in those that had been radish-based mixtures. Oats grain yield in the harvested crop was highest following the most biodiverse mixture (oats, *Phacelia*, vetch and radish) and lowest in the control plots.

The second experiment involved single species plots of radish, buckwheat, vetch, *Phacelia*, and oats and so provided a clearer picture of how different species contributed to any beneficial outcomes. The radish plots supported more epigeic earthworms than other cover crops species. In this experiment, the amount of leaf litter deposited on the ground was also recorded and found to be associated with earthworm abundance, with the highest earthworm numbers where leaf litter production and deposition were greatest. Epigeic earthworm species tend to be short-lived and reproduce relatively

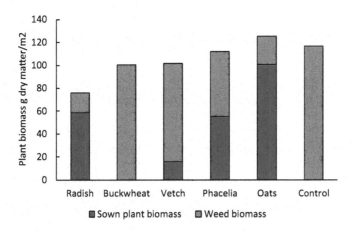

Figure 3.3 Late winter cover crop and weed biomass

rapidly and numbers can respond even to short-term changes such as the introduction of some cover crop species into the rotation.

As in the previous experiment, there was no difference in slug numbers between cover crop species or control plots, but the abundance of herbivorous nematodes was highest in *Phacelia* plots. Also in line with the previous multi-species experiment, there was a significant negative relationship between weed biomass and that of the sown cover crop, especially in radish and oats. In the following spring-sown oats crop, the yield of the harvested crop was highest, and weed biomass lowest, following the radish cover crop (Figure 3.3).

Yields were low across all plots because of difficult growing conditions in that year. Once all fixed and variable costs were accounted for, only oats following radish, or following oats were profitable, and only marginally. This highlights the practical and economic challenges of adopting spring crops and cover crops on clay soils. Establishment of crops in the spring can be difficult and delayed when conditions are wet, and early establishment and the correct species of preceding cover crop are essential if there is to be sufficient growth to deliver the anticipated benefits.

In this single species experiment there was also a study of soil phosphatase activity at the end of the cover crop period (March), and in the following oats crop (June) (Reynolds et al., 2017). Alkaline phosphatase is produced by microorganisms, whereas acid phosphatase is also produced by plants. These extracellular enzymes are deposited in the soil, either through cell turnover or secreted by organisms to utilise phosphate reserves within organic matter. There was no effect on alkaline phosphatase activity at the time of cover crop incorporation into the soil (March), but by the point of maturation of the following oats cash crop (June), significant differences were detected, with the greatest enzyme activity following an oats cover crop. Acid phosphatase activity showed significant species-related differences at both sampling dates, with the magnitude increasing by June. Again, plots following an oats cover crop showed the greatest activity, followed by *Phacelia.*   ·

This work indicates that soil phosphatase enzymes may be affected by the presence of a cover crop, that this effect is apparently species-dependent (not dependent on the amount of biomass from the cover crop) and that some cover crops could therefore be a potential means to enhance soil phosphorus cycling, reducing the need for application as fertiliser. However, an important caveat is that these results were not repeated in a second-year experiment, or in the lab, so need to be treated with caution.

Harvested crops remove phosphorus from the field, requiring application of phosphate fertiliser, but as a finite resource with depleted sources and increasing prices, there is a need to improve use efficiency of this input. Around 85% of applied phosphate remains tightly adsorbed to soil particles and is less available to crops (Johnston et al., 2014), but this store of legacy soil phosphate can still be a useful source of P when fertilisers become too expensive to apply. A recent lab-based experiment using soils from the Welland river basin suggests that, in the absence of phosphate fertiliser application, crop production could continue for several years on both calcareous and clay soils by drawing down from this pool of 'legacy' phosphorus in the soil before crop production becomes limited (RePhoKUs project, 2021). The extent to which this is possible is dependent on initial Olsen P status. An upper index 3 soil provides around 200 kg/ha, and an upper index 2 soil provides around 100 kg/ha, representing respectively approximately eight and four years of arable cropping. The phosphorus mobilised over this period represents only a small proportion of the total pool of legacy phosphorus. Like rock phosphate, this is a finite resource, but it may be possible to optimise cropping patterns, cover crops and soil biology to extend the period over which legacy P can be made available, enabling farmers to limit the purchase of phosphate fertiliser when global prices are high, or supplies are restricted. Such an approach, exploiting the biology of soils on individual farms to meet nutrient and other requirements, needs a more flexible approach to nutrient management than is currently prescribed.

Cultivation

Reducing the intensity of soil cultivation reduces soil disturbance and can enhance soil biology and function, including carbon sequestration (Cooper et al., 2021). From 2001, the Allerton Project's farm business made a gradual shift away from traditional plough-based cultivation systems, through reduced non-inversion tillage using discs and tines, to a direct drilling approach in which the seed of one crop is drilled directly into the stubble of the previous one with minimal disturbance of the soil. Throughout this period, various replicated plot experiments have contributed to a much better understanding of the relationship between soil management and soil function. Whereas the study by Hamad (2018) referred to earlier, had selected compacted areas of fields, in the experiments described here, plots are randomly replicated, making a direct comparison between plough and reduced tillage or direct drilling possible.

In a comparison of plough with a reduced tillage approach at Loddington, Cooper (2006) showed that reduced tillage reduced bulk density and increased organic matter. Surface roughness, including crop residues and plant cover, had a major impact on runoff and erosion rates, and soil loss from fields increased linearly in response to the runoff rate. However, runoff rates were less pronounced under both reduced cultivation and plough when storm events first started, and the differences between them increased as the storms progressed, with greater increases in runoff rate in the plough plots than in reduced cultivation. In the planar plots used for the experiments, the rates of soil loss ranged from 0.008 to 0.120 t/ha/yr, but at the field or landscape scale, we can expect erosion rates to be higher on more undulating slopes where flow is more concentrated.

The mean phosphorus concentration in runoff was significantly higher, at 0.8 mg/L, from ploughed plots than the 0.1 mg/L recorded from reduced tillage plots. Phosphorus, potassium and nitrogen concentrations in eroded sediment were all significantly higher than in soil within the plots from which it was derived and this enrichment was higher in sediment from plough plots than from reduced tillage ones.

Allton et al. (2007), working on the same plots, were able to show that the microbial community structure was affected by cultivation, with fungal community size in particular being greater under reduced tillage treatments than under ploughing. There was also associated higher summer moisture retention and a tendency for reduced soil erosion in the reduced tillage plots. Under controlled conditions in the lab at Cranfield University, sterilised soil from the plots was inoculated with microbial samples obtained from soils taken from each of the different tillage approaches adopted at Loddington. The propensity to runoff was lower, and infiltration rates were higher where soil had been inoculated with microbial material from the reduced tillage plots. This is likely to be a result of fungal hyphae and associated humic substances such as glomalin which improve the aggregate stability of the soil, maintaining porosity.

In another project, as well as comparing ploughed plots with reduced tillage ones, Deasy et al. (2008, 2010) also tested the effect of cultivation direction, and a low grassy 'beetle bank'[2] along the contour of the slope. The material lost in surface runoff from the various treatments was measured. Total carbon was mainly (58–99%) transported in dissolved rather than particulate form, but in this study, there was no difference in carbon loss between plough and reduced tillage plots. The average total phosphorus loss was 360 g/ha/y, with 58–95% being in particulate form. The total phosphorus concentration in runoff was 1.37 mg/L, but this was variable and tended to be higher, along with associated suspended sediment concentrations, in the early stages of storm events. Minimum tillage was generally an effective means of controlling sediment and phosphorus loss, but this effect differed between years and was most consistent in plots cultivated up and down the slope (34-62% reduction).

A beetle bank across the slope reduced runoff and losses of suspended sediment and total phosphorus by 32–45% in ploughed plots. Beetle banks were also beneficial in reduced tillage plots, but to a lesser extent. Tramlines (tractor wheelings) were found to be the major pathway for runoff from arable hillslopes, with nearly three times more from plots containing tramlines than from plots without tramlines. In general, sediment and phosphorus loss in runoff was 2–5 times higher from tramlines than it was from the rest of the field. Although reduced tillage, contour cultivation and beetle banks have the potential to be cost-effective mitigation options, the latter two are not practical in most fields with slopes in multiple directions; tramline management could be one of the most promising approaches for mitigating diffuse pollution of water from arable land.

Deasy et al. (2014) analysed the runoff data from the perspective of flood risk prevention. They concluded that while reduced tillage had the potential to reduce erosion, it may lead to increased surface runoff generation in a storm event, with greater storm event runoff peaks, and longer runoff responses to rainfall than for ploughing. However, this is counter to other studies (e.g. Strudley et al., 2008) which report increased macropore connectivity and associated infiltration rates. The differences could be explained by the fact that the plots at Loddington were relatively recently established with poorly developed soil structure.

As a result of this work, another plot-based project (Silgram et al., 2015), tested various approaches for reducing runoff and erosion along tramlines, including reprofiling the tramlines to shed water to the sides, the use of a tine to improve infiltration, reseeding to establish vegetative cover, surface disruption using a small rotary harrow, and the use of low ground pressure tyres. In one year, there was a significant effect of low ground pressure tyres on runoff, sediment concentrations and sediment loads, while in another the harrow treatment reduced sediment and phosphorus loss more, although there was also an additional benefit of applying low ground pressure tyres as well. However, the rotary harrow is sensitive to soil conditions on clay soils, being ineffective when the ground has dried hard, or is wet and highly plastic, whereas low ground pressure tyres are effective under a range of conditions, especially wet conditions when soil is most susceptible to compaction.

In February and March 2021, we investigated the relationship between soil compaction, water infiltration rates and soil organic carbon by gathering data and samples from 14 local fields where there were clear signs of compaction affecting crop performance. We identified compacted areas and relatively uncompacted areas in each field and gathered data from each. Penetration resistance data confirmed that areas identified as compacted were indeed significantly more so than the comparison sites, and infiltration rates in compacted areas were close to zero, whereas in less compacted areas they were much more variable. Across the relatively uncompacted sites, infiltration rates were significantly correlated with penetration resistance to 20 cm depth, and with soil organic matter (Figure 3.4). A 1% increase in soil organic matter was associated with an increase in water infiltration rate of about 3 mm/minute. This illustrates the value of organic matter in increasing infiltration and reducing surface runoff with the associated implications for managing catchment scale flood peaks and sedimentation of drainage channels, as well as for water quality objectives. Soil organic matter was highly negatively correlated with bulk density, illustrating the importance of organic matter in reducing soil compaction. Estimates of carbon stocks in these arable fields ranged from 37 to 125 tonnes/ha with an average of 58 t/ha.

While soil compaction can be a major issue in plough-based systems and is one that has increased in recent decades as a result of reduced organic matter, larger machinery, and wetter autumns, compaction can also occur in direct-drilled fields, especially on clay soils and where drainage is poor. This can be alleviated by ploughing periodically in the crop rotation, or when compaction occurs, but this may reverse any soil health benefits achieved through the previous adoption of direct drilling. The use of a low disturbance subsoiler to break up compacted soil at depth, without disturbing the soil surface or otherwise inverting the soil, maybe a pragmatic compromise.

These two approaches have been compared experimentally over a two-year period at Loddington (Bussell et al., 2021). For this experiment, the entire area of the plots was deliberately compacted by driving a tractor and loaded trailer uniformly across the plots. The plots were then either ploughed or sub-soiled for comparison with control plots which were drilled without any mitigation. The two approaches were also compared with the use of mycorrhizal fungal inoculation of the seed at drilling. This latter

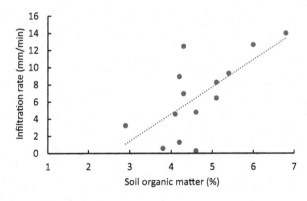

Figure 3.4 Soil organic matter and water infiltration rates for 14 local fields

approach has no physical effect on the soil, but could potentially improve the crop's access to moisture and nutrients through the action of fungal hyphae associated with the crop roots.

As expected, measures of compaction (visual assessment, bulk density and penetration resistance) were higher in the compacted direct-drilled control and mycorrhizal inoculant plots than in the plots where physical steps had been taken to reduce compaction by ploughing or subsoiling. The largest differences were between 5 and 25 cm in the soil profile. Earthworm numbers were higher in the unalleviated direct-drilled and inoculated plots than in ploughed and sub-soiled plots, although this was only statistically significant in the first year when the field was drilled with barley. In the second year, when the field was growing field beans, earthworm numbers were generally lower across all plots, probably because of drier conditions in that year.

The most interesting results to come out of this experiment were for greenhouse gas flux from the soil which was measured using a Gasmet gas analyser (Bussell et al., 2021). These gases comprise CO_2 and N_2O, of which the latter has nearly 300 times the global warming potential of CO_2 (IPCC, 2007). As discussed earlier, CO_2 is a measure of soil fauna respiration. It was overall around 130% higher in the summer when biological activity increases, but in the winter, was significantly higher in plots where the soil had been disturbed than in the undisturbed plots. N_2O flux, on the other hand, was higher in the undisturbed plots than the ploughed or sub-soiled ones in the winter. N_2O is produced principally by denitrifying bacteria which are associated with anaerobic conditions such as those found in wet compacted soils. Although the amounts of N_2O recorded were considerably lower than CO_2, their high global warming potential meant that the overall impact of the two gases was similar in terms of climate change implications (Figure 3.5). At least in these compacted conditions, climate change implications of ploughed, sub-soiled, inoculated and undisturbed direct-drilled plots were equivalent, during winter months. Once the lower CO_2 emissions recorded in direct-drilled treatments over summer, and lower diesel use for crop establishment in the direct-drilled treatments are factored in, the net effect is that even the compacted direct-drilled plots have a lower climate change impact than either the plough or sub-soiled plots. In situations where physical compaction alleviation is necessary, subsoiling has the least environmental impact.

Crop yields in this experiment did not differ significantly between treatments in the barley or beans, but in the barley, showed a trend towards being higher in plough and sub-soiled plots than in inoculated and control plots (without physical compaction alleviation). The economic impact of subsoiling was positive compared to the control due to an improvement in yield when compaction was alleviated. Subsoiling also performed marginally better than plough, despite lower yields because of the cost of additional cultivations required after ploughing. This implies that, instead of ploughing,

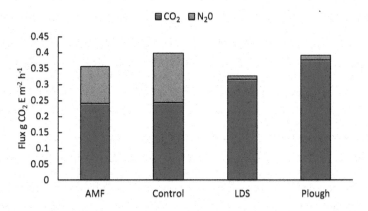

Figure 3.5 Carbon dioxide equivalent emissions for CO_2 and N_2O in relation to compaction alleviation treatments and control plots. AMF = Arbuscular Mycorrhizal Fungal inoculant, LDS = Low Disturbance Subsoiler.

subsoiling can be adopted to reduce compaction in direct-drilled fields without economic penalty while retaining the soil health benefits associated with a direct drill approach. This approach to crop establishment is estimated to applicable to 25% of the European agricultural area (van Delden, 2021).

In another experiment (Bussell, 2021a), we investigated in more detail the implications for soil properties and crop performance of ploughing long-term direct-drilled land. As with the previous experiment, this one was inspired by farmers' concerns about potential negative impacts, should they feel it necessary to plough at some stage to control weeds or reduce compaction in a direct-drilled system. This study focused on soil biology, and specifically on what aspects of soil biology are indicative of soil health. We ploughed out three strips in a field that had been direct-drilled for seven years, creating three ploughed and three direct-drilled plots. We then repeated the ploughing in the two subsequent years while continuing with the direct drilling in the other plots.

In the first year (2018) we found smaller numbers of earthworms in the ploughed plots. This was mostly due to a decline in endogeic earthworms, which are most at risk from direct plough damage. Topsoil structure (25 cm depth) was monitored using Visual Evaluation of Soil Structure (VESS, Ball et al., 2007). After three consecutive years, we found a higher VESS score (poorer structure) in the ploughed plots, while in the direct-drilled plots, the structure in the top layer had a better structure, in contrast to the ploughed soil where the soil aggregates were larger with less continuous pore space.

Cultivation is known to reduce survival and abundance of some beneficial invertebrates such as carabid beetles which contribute to crop pest control,[3] with impacts varying between species and being expressed as change in invertebrate community composition rather than overall abundance

(Holland, 2004; Holland & Luff, 2000). In Year 3 of the experiment, we used emergence traps (one in each plot), primarily to record the invertebrates emerging from pupae in the soil during the March–May period as in earlier work, we found greater numbers of two carabid species (*Notiophilus biguttatus* and *Bembidion guttula*) in reduced tillage than in ploughed fields (SOWAP, 2007). However, in the more recent experiment, we found no difference in abundance of Carabid or Staphylinid beetles and abundance was consistent across ploughed and direct-drilled plots, with three peaks in emergence during the months of April and May (Szczur, 2020). Abundance of Diptera, represented mainly by Chironomidae (non-biting midges) and Sciaridae (dark-winged fungus gnats) was significantly higher in the direct-drilled plots. Parasitic wasps, many of which are parasitoids of aphids did not differ between ploughed and direct-drilled plots. Their emergence occurred from late April to late May, a period in which aphids can be actively colonising and feeding on crops (Figure 3.6).

Soil-dwelling Collembola (springtails) were also trapped in the emergence traps with significantly higher numbers in direct-drilled plots through March and most of April. Although Collembola are small, their contribution to mesofauna biomass is low but they play an important role in the decomposer food web consuming decaying plant material and other organisms (Crotty, 2021). As a result, they have a large effect on soil structure and composition, and the release of nutrients through microbial action on their faecal material.

In contrast to our experiment exploring alleviation of deliberately created compaction in direct-drilled plots, we found no difference in N_2O emissions between ploughed and direct-drilled plots by the time the latter were in their tenth year. This is an important finding. It suggests that while N_2O emissions can be higher under direct drilling than ploughing where the soil is compacted and waterlogged, given time, the restructuring of the soil

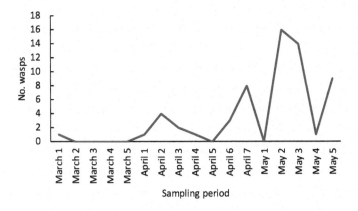

Figure 3.6 Parasitic wasp emergence into emergence traps by trapping periods from March to May 2020 (Szczur, 2020)

through biological activity creates aerobic conditions that are less conducive to the bacterial denitrification process that releases N_2O.

Bussell (2021a) explored this relationship between microbial activity and greenhouse gas emissions further. She took five samples from each plot in April, May, July and September for analysis using MicroResp[TM] (James Hutton Institute). This technique uses a range of organic acids, amino acids and sugars that are commonly found in soils as substrates. The average total respiration (μg CO_2-C g^{-1} soil h^{-1}) of soil to all substrates added provides an Average Metabolic Response (AMR), giving a measure of microbial activity. Summing the positive responses to substrate additions provides a Community Metabolic Diversity (CMD). This is a simple measure of ability of the soil microbial community to effectively metabolise the range of substrates, and therefore a measure of functional diversity. We also collected data on CO_2 and N_2O emissions from the same sites as the soil samples were taken, and at the same sampling times in order to capture in-field data on emissions using the Gasmet gas analyser (Figure 3.7).

AMR and CMD were both higher in direct-drilled plots than ploughed plots and increased from April to July, declining again in September. CO_2 flux was significantly higher in direct-drilled plots than in ploughed ones.

Figure 3.7 Gasmet gas analyser in experimental plots, with emergence traps in the background

N_2O flux was extremely low in both treatments, with no significant difference between treatments. This reinforces the finding that where the soil is well structured and not compacted, direct drilling is not associated with higher N_2O emissions. This is important given the high global warming potential of N_2O and counter to the concerns arising from earlier studies (Mangalassery et al., 2014; Palm et al., 2014), but is in line with those of Cooper et al. (2021).

Across all six plots, CO_2 flux increased significantly with CMD, but higher CMD in the direct-drilled plots was also associated with lower N_2O flux. There was a similar response for AMR, but with only a weak trend for N_2O. Soil organic matter was approximately 0.5% higher in direct-drilled plots than ploughed plots after three years, and there was a positive relationship between organic matter and both CMD and AMR. Higher organic matter was also associated with higher CO_2 flux, and lower N_2O flux. Examination of the levels of labile (active) active carbon in the top 10 cm of the soil revealed that these were higher in the direct-drilled plots than in the ploughed plots (Bussell, 2021b). Taken together, these results suggest that the biologically active direct-drilled soils are capturing and actively cycling carbon, as well as sequestering carbon, faster than the ploughed land.

These results indicate that the direct-drilled plots were associated with higher microbial activity, higher microbial functional diversity, higher active cycling of carbon, and higher more stable organic carbon than the ploughed plots. Given the importance of the microbial community in nutrient cycling, this suggests that availability of nutrients to crops may be higher in direct-drilled plots, with higher efficiency and resilience to environmental shocks associated with the enhanced functional diversity. This may include mobilisation of legacy phosphorus, reducing the need for application of phosphate fertiliser, as discussed earlier in this chapter.

Grass leys

Another approach that reduces cultivation and provides soil cover is the adoption of grass leys within the arable rotation. These provide soil stability, normally for three or four years, and roots improve soil structure and organic matter. An experiment testing a range of modern deep-rooting grass ley cultivars (Stoate et al., 2021) has contributed to our understanding of these and related issues. This approach to integration of grass into arable rotations is estimated to be applicable to 38% of the European agricultural area (van Delden, 2021).

The test grasses comprised four modern Festulolium cultivars and an agricultural cocksfoot. Festuloliums are four-way hybrids between perennial ryegrass (*Lolium perenne*), Italian ryegrass (*L. multiflorum*), meadow fescue (*Festuca pratense*) and tall fescue (*F. arundinacea*), all developed for their deep-rooting properties and value as forage for ruminants. Longevity, forage quality, and root architecture are all determined by the resulting genotype of each cultivar. The cultivars were developed largely to address the challenges presented to livestock farmers by increasing drought conditions in various parts of the world, but their deep-rooting properties could also potentially contribute to improved water infiltration rates, with implications for flood risk management and water quality, and to carbon sequestration at depth in the soil profile. MacLeod et al. (2013), working with another Festulolium, 'Prior', reported significantly reduced runoff in 29 out of 33 rainfall events, compared to perennial ryegrass and meadow fescue.

In our experiment, the test cultivars formed a 50% component of the grass seed mix, with the remaining 50% comprising a standard ryegrass and clover mix. Control plots consisted of the standard mix alone. Over the four-year life of the plots, grass was cut for silage each year and the aftermath was grazed by weaned lambs. In all but the last year, store lambs grazed the plots in the spring. Our grazing observations revealed that lambs grazed treatment plots at least as much as the plots containing Festulolium or cocksfoot, suggesting that there was no negative effect in terms of palatability of the introduced cultivars. Yield and quality of the sward were also similar across the various test cultivars and control plots.

Water infiltration rates (measured using a double-ring infiltrometer) were significantly higher in plots containing the Festulolium cultivars, Fojtan and Lofa, and the cocksfoot, Donata in the first year (Figure 3.8), with Fojtan infiltration being three times higher than the control, demonstrating potential benefits for meeting catchment management objectives. However, this was not repeated in the following year. Our results, combined with direct observation of the plots, caused us to think that a combination of harvesting for silage and by grazing sheep, and associated compaction, could be having direct and indirect effects on subsequent water infiltration rates. We know, for example, that harvesting of above-ground biomass restricts below-ground biomass in the form of root volume (Dawson et al., 2000).

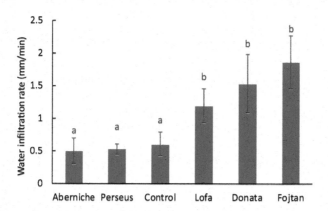

Figure 3.8 Water infiltration rates for four Festuloilums and one Cocksfoot cultivar in relation to standard ryegrass/clover plots (control)

As a result, a section of the plots was fenced off to exclude harvesting for silage or grazing by lambs in Years 3 and 4 to provide a comparison with the adjacent section of the plots where both these activities continued as before.

Data collection in Year 3 in the fenced and unfenced areas was confined to Donata, Fojtan and control plots (Stoate et al., 2021). This revealed that compaction, as measured by penetration resistance, was higher in the unfenced cut and grazed areas than in the fenced areas, confirming our suspicion that grazing and silage cutting were causing compaction. Curiously, root volume at 70 cm depths in unfenced areas was higher for the standard ley mixture than for the test cultivars, suggesting that silage cutting and grazing activity may limit root growth in Festuloliums and cocksfoot more than in standard ryegrass clover leys. However, in the unharvested areas, Fojtan had significantly and substantially higher root volume at 70 cm. In Year 4, root volume was measured in all plots. Four of the five test cultivars had higher root volume in the fenced off area than in the cut and grazed area, with Fojtan being the cultivar with the highest root volume at 70 cm (62% greater than the control in fenced areas), but this difference was not statistically significant. MacLeod et al. (2013) showed that Festulolium root growth was highest early in the life of the plant and that subsequent root senescence was responsible for improving soil porosity.

Soil organic carbon did not differ between treatments in our study but was significantly higher at 15 cm and 40 cm depths than at 70 cm, in line with the results referred to earlier for other land uses (Hamad, 2018). Data collection in the grass plots in Year 2 by Clarke (2019) explored this further with detailed analysis of a range of carbon fractions. The proportion of recalcitrant (stable) and labile forms of carbon influences the extent to which soil organic carbon contributes to genuine sequestration. Labile carbon is important for microbial activity and associated nutrient cycling and crop performance, as discussed above, but recalcitrant forms are necessary for sequestration. Carbon sequestered in soil through the adoption of grass leys and other management practices can also easily be lost again when the land is returned to arable cropping unless it is stored below the plough layer (around 25–30 cm depth).

In Clarke's (2019) study, soil samples from soil core segments were weighed and heated in a furnace for three consecutive three-hour cycles (at 200°C, 325°C and 550°C) and were weighed again after each cycle to obtain soil organic matter content. This stepwise process allowed the loss on ignition to be measured at each stage, providing an estimate of the proportion of labile and recalcitrant forms of carbon for each depth. There were significant reductions in the relatively labile organic matter fractions with increasing depth. The most recalcitrant form showed no significant pattern in relation to soil depth, while dissolved organic matter was the only measured organic matter fraction which significantly increased with depth suggesting that deeper soils contained more dissolved organic carbon. Perhaps as interesting as the significant trends was the fact that the most recalcitrant carbon was observed consistently throughout the soil profile. The fact that more labile carbon dominated the upper soil horizons, while deeper soil was dominated by more persistent carbon was consistent with previous research (Cambardella, 2005; Rumpel & Kogel-Knabner, 2011). The same research has also found that a key driver for the input of organic carbon in deep soils is typically through the deposition of dissolved organic carbon through preferential flow pathways (Rumpel & Kogel-Knabner, 2011).

Subsequent research by Holmes et al. (2021) in the same plots revealed that recalcitrant fractions of carbon were highest in Fojtan plots and lowest in plots of Perseus and Aberniche. Fojtan is genetically closest to the stress-resistant fescues which produce more complex long-chain defensive compounds which are harder to break down and are represented in their root exudates (Buta & Spaulding, 1989). These would be more recalcitrant than simple short-chain sugars which are produced by Lolium-dominated cultivars.

Our research suggests that deep-rooting grass cultivars such as Fojtan can therefore contribute to societal objectives for both water infiltration and carbon sequestration below the plough layer so long as management of the sward does not restrict root growth. To be realised, these societal benefits may be dependent on restricting harvesting intensity through approaches such as mob grazing or agri-environment scheme incentives. As autumns are becoming wetter, it may also be necessary to restrict the timing of grazing to avoid this period so as to avoid shallow compaction.

Further increasing botanical diversity in grass leys by incorporating additional grasses, legumes and other broad-leaved plants may further improve soil function. Early results from an experiment comparing a 'herbal ley' with a conventional grass/clover ley suggest that soil organic carbon increases more rapidly in the former than the latter, with higher organic matter and water-holding capacity in both types of ley (Leake, 2021).

Our research suggests that grass leys comprising carefully selected species and cultivars could make a valuable contribution to improvement of soil health in arable rotations, complementing the management practices we have studied for annual arable crops. The reduced soil disturbance associated with leys complements that associated with direct drilling or reduced tillage, accompanied by an appropriate compaction alleviation method where this is needed. Enhanced soil health in the form of wide-ranging soil biota contributes to soil functions such as water infiltration, carbon sequestration and nutrient cycling, including possible mobilisation of legacy phosphorus, as well as contributing to the ecology of the wider farmland environment.

Notes

1 Compaction and infiltration in grassland are discussed later in the chapter.
2 Beetle banks are described in Chapter 5.
3 Beneficial predatory beetles are discussed in Chapter 5.

References

Ball, B. C., Batey, T. & Munkholm, L. J. (2007) Field assessment of soil structural quality – a development of the Peerlkamp test. *Soil Use and Management* 23, 329–337. DOI:10.1111/j.1475-2743.2007.00102.x

Bussell, J., Crotty, F. & Stoate, C. (2021) Comparison of compaction alleviation methods on soil health and greenhouse gas emissions. *Land* 10, 1397. DOI:10.3390/land10121397

Bussell, J. (2021a) *Soil Biology and Soil Health Partnership Project 12: Testing Soil Health and Resilience Using Soil Respiration Activity.* Project Report No. 91140002-12. Agriculture and Horticultural Development Board.

Bussell, J. (2021b) Unpublished data.

Buta, J. G. & Spaulding, D. W. (1989) Allelochemicals in tall fescue-abscisic and phenolic acids. *Journal of Chemical Ecology* 15 (5), 1629–1636.

Cambradella, C. (2005) Carbon Cycle in Soils: Formation and Composition. In: *Encyclopaedia of Soils in the Environment.* Academic Press, Cambridge, MA. 170–175.

Clarke, J. (2019) *Examining the impacts of plant species diversity and rooting characteristics on the accumulation of soil organic matter in temporary grasslands.* Unpublished MRes thesis. University of York.

Cooper, S. E. (2006) *The role of conservation soil management on soil and water protection at different spatial scales.* Unpublished PhD thesis. Cranfield University.

Cooper, H., Sjögersten, S., Lark, R. M. & Mooney, S. (2021) To till or not to till in a temperate ecosystem? Implications for climate change mitigation. *Environmental Research Letters* 16 (5) DOI:10.1088/1748–9326/abe74e

Crotty, F. & Stoate, C. (2019) The legacy of cover crops on the soil habitat and ecosystem services in a heavy clay, minimum tillage rotation. *Food and Energy Security.* DOI:10.1002/fes3.169

Crotty, F. (2021) Soil Organisms Within Arable Habitats. In: Hurford, C., Wilson, P. & Storkey, J. (Eds.) *The Changing Status of Arable Habitats in Europe.* Springer Nature, Cham. 123–138.

Dawson, L. A., Grayston, S. J. & Paterson, E. (2000) Effects of Grazing on the Roots and Rhizosphere of Grasses. In: Lemaire, G., Hodgson, J., de Moraes, A., Nabinger, C. & de Carvalho, F. (Eds.) *Grassland Ecophysiology and Grazing Ecology.* CABI, Wallingford. 61–84.

Deasy, C., Quinton, J. N., Silgram, M., Jackson, R. J., Bailey, A. P. & Stevens, C.J. (2008) *Field Testing of Mitigation Options (Final Report to Defra for Contract PE0206).* Lancaster University & Defra, Lancaster and London.

Deasy, C., Quinton, J. N., Silgram, M., Bailey, A. P., Jackson, B. & Stevens, C. J. (2010) Contributing understanding of mitigation options for phosphorus and sediment to a review of the efficacy of contemporary agricultural stewardship measures. *Agricultural Systems* 103, 105–109.

Deasy, C., Titman, A. & Quinton, J, N. (2014) Measurement of flood peak effects as a result of soil and land management, with focus on experimental issues and scale. *Journal of Environmental Management* 132, 304–312.

Hamad, F. D. (2018) *The consequences of land management, particularly compaction, on soil ecosystems.* Unpublished PhD thesis. University of Leicester.

Holland, J. M. & Luff, M. L. (2000) The effects of agricultural practices on Carabidae in temperate agroecosystems. *Integrated Pest Management Reviews* 5, 109–129.

Holland, J. M. (2004) The environmental consequences of adopting conservation tillage in Europe: reviewing the evidence. *Agriculture, Ecosystems and Environment* 103, 1–25.

Holmes, J., et al. (2021) Unpublished data.

IPCC (2007) *Fourth Assessment Report*. Intergovernmental Panel on Climate Change.

Johnston, A. E., Poulton, P. R., Fixen, P. E. & Curtin, D. (2014). Phosphorus: Its efficient use in agriculture. *Advances in Agronomy* 123, 177–228.

Leake, J. (2021) Unpublished data

Lee, M., Stoate, C., Crotty, F. et al. (2017) *Defra Sustainable Intensification research Platform*. SIP Project 1: Integrated Farm Management for Improved Economic, Environmental and Social Performance (LM0201). Objective 1.2: Identify and Develop Farm Management Interventions for Sustainably Intensive Agriculture. Defra, London.

MacLeod, C., Humphreys, M. W., Whalley, W. R., Turner, L., Binley, A., Watts, C.W., Skøt, L., Joynes, A., Hawkins, S., King, I. P., O'Donovan, S. & Haygarth, P. M. (2013) A novel grass hybrid to reduce flood generation in temperate regions. *Scientific Reports* 3, 1683.

Mangalassery, S., Sjögersten, S., Sparkes, D. L., Sturrock, C. J., Craigon, J. & Mooney, S. J. (2014) To what extent can zero tillage lead to a reduction in greenhouse gas emissions from temperate soils? *Scientific Reports* 4, 1–8.

Palm, C., Blanco-Canqui, H., DeClerck, F., Gatere, L. & Grace, P. (2014) Conservation agriculture and ecosystem services: an overview. *Agriculture, Ecosystems & Environment* 187, 87–105.

RePhoKUs project (2021) Unpublished data.

Reynolds, S., Ritz, K., Crotty, F., Stoate, C., West, H. & Neal, A. (2017) Effects of cover crops on phosphatase activity in a clay arable soil in the UK. *Aspects of Applied Biology* 136, 215–220.

Rumpel, C. & Kogel-Knabner, I. (2011) Deep soil organic matter: a key but poorly understood component of terrestrial C cycle. *Plant and Soil* 338, 143–158.

Sheppard, D. (2014) *Earthworms in England: Distribution, Abundance and Habitats. Natural England Commissioned Report NECR145*. Natural England, Exeter.

Silgram, M., Jackson, B., McKenzie, B., Quinton, J., Williams, D., Harris, D., Lee, D., Wright, P., Shanahan, P. & Zhang, Y. (2015) *Reducing the Risks Associated with Autumn Wheeling of Combinable Crops to Mitigate Runoff and Diffuse Pollution: A Field and Catchment Scale Evaluation*. AHDB Project Report no. 559.

SOWAP (2007) *Soil and Surface Water Protection Using Conservation Tillage in Northern and Central Europe (SOWAP)*. Technical Final Report. LIFE03 ENV/UK/000617. EU LIFE.

Stoate, C., Bussell, J. & Fox, G. (2021) Potential of Deep-Rooting Agricultural Grass Cultivars for Increasing Water Infiltration and Soil Organic Carbon. In: Aubertöt, Bertelsen et al. (eds) *Intercropping for Sustainability: Research Developments and Their Application. Aspects of Applied Biology*. 146. Association of Applied Biologists, Warwick.

Strudley, M. W., Green, T. R. & Ascough, J. C. (2008) Tillage effects on soil hydraulic properties in space and time: state of the science. *Soil and Tillage Research* 99, 4–48.

Szczur, J. (2020) Unpublished data.

van Delden, H. (2021) SoilCare project unpublished data.

4 Wildlife and our food

In these days of sterile packaged food, it is easy to forget that our food production is inextricably linked with the lives of other species. Some of those species live in or on the food we store in our homes or on farms. Others are associated with food as it is growing, either as pests, parasites, competitive weeds or as species that simply live alongside the plants or animals that provide our food. Some of the latter are iconic species that are associated with farmland and have a strong cultural value attached to them. There are other species which have direct benefits to the production of our food, for example through the control of pests or the pollination of crops. Species living in the soil have multiple beneficial roles, as described in Chapter 3.

The plants we know as weeds have evolved alongside the development of arable farming over thousands of years (Stoate, 2011). Species composition of weed communities has changed over that time in response to changes in management practices, including those intended to control weeds, and to differences in climate and soil type. Some formerly abundant species have been lost as a result of mechanical and chemical weed control and now represent some of our rarest plants (Stoate & Wilson, 2021). The number of plant species present on clay soils such as those at Loddington is lower than on lighter land where there has been a long history of cultivation. Field madder *Sherardia arvensis*, Long-headed poppy *Papaver dubium* and common cudweed *Filago vulgaris* are the scarcest of the arable weeds to have been recorded on the farm at Loddington.

Other species have become more dominant, with black-grass *Alopecuris myosuroides* being a highly competitive and adaptable species that is currently a major influence on crop rotations, especially with recent evidence of evolved resistance to a range of herbicide strategies (Comont et al., 2020). As a mainly autumn-germinating species, black-grass has benefited from the switch from spring cropping to autumn-sown crops that has taken place since the 1960s. Simpler rotations have been associated with a greater reliance on herbicides for weed control.

In grassland, a combination of increased fertiliser use and reseeding of semi-natural swards with grass cultivars that have been developed to respond to those fertilisers has led to loss of botanical diversity. Botanically

DOI: 10.4324/9781003160137-4

diverse meadows are now scarce in the area around Loddington, as they are more widely across the UK and much of Europe. One field at Loddington continues to be managed as a traditional hay meadow to maintain the range of plant species that are characteristic of this habitat. These include now scarce or localised species such as adders' tongue fern *Ophioglossum vulgatum*, bugle *Ajuga reptans*, harebell *Campanula rotundifolia*, tormentil *Potentilla erecta*, greater burnet-saxifrage *Pimpinella major* and pignut *Conopodium majus*. The latter supports a colony of chimney sweeper moths *Odezia atrata*, a day-flying species for which pignut is the larval foodplant.

Other invertebrates are more problematic. In grassland, mud snails *Galba truncatula* are intermediate hosts of liver fluke *Fasicola hepatica* which are internal parasites of sheep and cattle and can cause considerable losses in production, including sudden deaths of sheep in early winter. The snails and fluke live in wet muddy areas around ponds, ditches, leaking water troughs and other wet areas that are trampled by livestock. Fluke cysts on grass are eaten by sheep or cattle and immature fluke make their way to the liver and ultimately inhabit the bile ducts from where the adults release eggs into the intestine. Climate change is resulting in a longer period of infection as mild wet conditions predominate, and eastern and northern areas of the country are increasingly affected.

Intestinal worms have a similar impact on sheep health and productivity. Newly hatched larvae live in the dung of sheep and cattle and third-stage larvae migrate to herbage where they are ingested by grazing animals. They become adults within around 14 days and start producing eggs soon after that, repeating the cycle. *Nematodrius* species have a longer, slower life cycle and their movement to herbage for ingestion by animals is triggered by temperature. Large numbers of larvae can therefore become available to animals in a concentrated period, even though they may have been deposited and hatched over a much longer period.

Natural resistance to worms (but not to fluke) develops in older ewes. Various anthelmintic treatments have been developed to treat ruminants for intestinal worms and liver fluke, but the parasites' resistance to most of them is becoming increasingly widespread. This requires a change in management involving a better understanding of parasite ecology and more judicious combinations of anthelmintic use, and the use of complementary measures.[1] Anthelmintics deposited in dung are known to be toxic to grassland invertebrates, most notably multiple dung beetle species which play an important role in the breakdown of dung and provide food for many vertebrates (Finch et al., 2020; Manning et al., 2017).

A survey at Loddington in 2017 revealed a total of 71 dung-living beetle species, 11 of which were locally scarce. These records provide an insight into beetle communities associated with livestock on the farm. The species include *Philonthus spinipes*, *Atholus bimaculatus*, *Cilia silphoides* and *Onthophagus coenobite* which feed directly on dung, but also *Carcinops pumillo* which feeds on fly eggs that have been laid in dung (Graham Finch, unpublished data, 2017).

Orthoptera (grasshoppers and crickets) are now an established component of the grassland invertebrate fauna at Loddington. The farm was colonised by lesser marsh grasshoppers *Chorthippus parallelus* in 1997. Roesel's bush cricket *Metrioptera roeselii* colonised the farm in 2002 and both species rapidly became widely distributed and abundant. A genetic study of lesser marsh grasshopper in 2002 found that there was as much genetic variation in the Loddington population as there was in samples from the wider countryside, including sites outside central England (Stoate, 2004). This suggests a rapid colonisation of the farm from far afield, consistent with the north-westwards range expansion of this and other species across the UK in response to climate change. At the same time, a simple feeding trial in which several grass species were offered to captive grasshoppers consistently identified *Holcus lanatus* and *Alopecurus pratensis* as the species to be consumed first. The combination of climate change and areas in which these grass species are present has therefore contributed to the colonisation of the farm by lesser marsh grasshoppers.

Arable fields also support a wide range of invertebrates. *Deroceras reticulatum* is the most common of the slug species at Loddington, often reaching high densities. Crotty (2017) recorded a density of 172 slugs m^{-2} (1.7 million slugs per hectare) as part of an experiment on cover crops. Brassica cover crops have a reputation for supporting higher numbers of slugs, but there was no evidence to support that in our study. As moisture is paramount for slug survival and growth, clay soils suit them well, and this is one of the animals that can benefit from reduced intensity of cultivation. Slugs cause major damage to germinating seeds and young plants, sometimes resulting in complete crop failure, and are particularly associated with oilseed rape *Brassica napus* and the following crop in the rotation. Where direct drilling leaves drill slots open, this creates moist conditions and easy access for slugs to move along the rows. Use of the molluscicide Methiocarb was banned because of impacts on other invertebrates and on vertebrates, while Metaldehyde which was used subsequently caused problems for drinking water supply. This stimulated water companies to employ teams of catchment advisors to introduce initiatives to enable farmers to reduce the movement of Metaldehyde to watercourses. Metaldehyde was withdrawn from use in 2021, leaving Ferric Phosphate as the main product for the control of slugs in arable crops.

Aphids are vectors of the cereal disease, Barley Yellow Dwarf Virus (BYDV), as well as causing direct harm to crops when abundant. As there is no chemical control of BYDV, control of aphids during the infective phase in the late summer and autumn is important. Spiders have a role to play in this as vast numbers of money spiders (Lynyphiidae) can be present at this time of year. Where stubble from the previous crop is left undisturbed, and not ploughed in, this supports the construction of webs and can potentially result in higher spider numbers, lower aphid numbers and lower incidence of BYDV, but this is not consistent from year to year (Holland, 2019).

Broad-spectrum insecticides are therefore widely used for aphid control, both for controlling BYDV and for reducing their direct impact on crops contributing to declines in insect numbers in arable systems, including those that are beneficial as pollinators or predators of crop pests. Neonicotinoid insecticides provide a much-publicised example. Even before these neonicotinoids were banned outright in the EU and the UK in 2018, resistance of aphids to them was an increasing concern.

The neonicotinoid ban resulted in an increase in pyrethroid application against cabbage stem flea beetle *Psylliodes chrysocephala*, a major oilseed rape pest, accelerating the development of flea beetles' resistance to pyrethroids (Willis et al., 2020). Flea beetles emerge from the soil in a rape crop and disperse in August and September to lay their eggs at the base of seedlings of the next year's crop. The larvae burrow into and through the growing plant during the winter and spring, often causing considerable damage to the stem, before emerging in May and June to pupate in the soil. The declining ability to control this pest has resulted in a dramatic decline in the area of oilseed rape grown, and therefore vegetable oil production in the UK.[2]

There is a similar story for bruchid beetles *Bruchus rufimanus* which are pests of field beans *Vicia faba*. Adult beetles feed on the pollen of bean flowers and lay their eggs on the newly forming pods. The larvae enter the beans and feed on them from the inside, where they pupate. The newly emerged adults then tunnel out of the bean, leaving a prominent hole. High-quality field beans are sold for human consumption, mainly in the Middle East, but crops are rejected and sold for animal feed at a lower price if there is evidence of damage caused by bruchid beetles. This represents a substantial economic loss, the incidence of which is increasing as bruchid beetles develop resistance to pyrethroid insecticides.

Invertebrates as pest predators and pollinators

Some invertebrate pests are controlled by predatory invertebrates such as spiders and Carabid and Staphylinid beetles. Most of these tend to be associated with field margin habitats where the vegetation structure and type influence the characteristics of the invertebrate predator community, although some can survive in the field year-round if the soil remains undisturbed (Chapter 3). Web-building spiders, mainly *Lepthyphantes tenuis*, are most abundant at Loddington where there is tall herbaceous vegetation (Haughton et al., 2000). Dense perennial grass sward provides overwintering habitat for many predatory beetle species. These move out into the crop in the spring where they eat aphids, the eggs and immature stages of slugs, and other crop pests, the distance travelled into the crop varying between species (Collins et al., 2002, 2003).

Woody field margin habitats are also occupied by predatory beetles. In 2013, we used pitfall traps to compare Carabid beetle abundance in two field margin hedges with that in two woodland edges, paired within two fields (Thompson, 2013). Carabid beetles were significantly (2.5 times) more abundant in woodland edges than hedges. A slightly higher number of species (24) was recorded in woodland edges than in hedges (20). Because pitfall traps are a measure of activity as much as of abundance, and as many Carabid species are highly mobile, emergence traps were also used in the two field margin habitats to assess the emergence of adult beetles from hibernation or pupae in each habitat. While numbers of beetles sampled were low, these results were generally consistent with those from pitfall trapping. Pitfall trap data also revealed significantly higher numbers of 'other beetles' and 'other invertebrates' in woodland edges, but significantly higher numbers of spiders in hedges than in woodland edges.[3] This diversity in beneficial invertebrate communities associated with different field margin habitats is likely to increase the capacity for control of crop pests at the field scale, depending on the distance travelled by different species into the crop.

The numerous Carabid, Staphylinid and other beetles present on farmland have a wide range of diets, including plants, fungi and other invertebrates. For example, many Carabid beetles feed on the seeds of weeds such as field pansy *Viola arvensis* and chickweed *Stellaria media* and by using exclusion cages to restrict larger animals' access to trays of seeds in 1999, we estimated that Carabid beetles were responsible for 50–90% of seed consumption (Tooley, 1999). Long-term monitoring of arable invertebrates at Loddington shows that numbers of Carabid and Staphylinid beetles have been declining (Figure 4.1). This is consistent with monitoring in other areas and may, in part, be due to the use of fungicides in crops as some species are fungal feeders (Ewald et al., 2016). Both in Sussex (southern England) and at Loddington, there has been an increase in numbers of some larger species, and competition with smaller species may partially explain the decline in numbers of the latter (Holland, 2021).

Ladybirds, lacewings and some hoverflies are obligate predators of aphids. Annual surveys of arable invertebrates at Loddington reveal that numbers of these predators and their prey fluctuate considerably from year to year. While hoverfly numbers have generally been higher in recent years than at the start of the project 30 years ago, ladybirds have declined substantially over that period (Figure 4.1). This is partially attributed to the population expansion of predatory harlequin ladybirds *Harmonia axyridis* which have colonised the UK in recent years and contributed to a reduction in the abundance of the smaller indigenous species (Brown & Roy, 2018).

Field margins provide an important habitat for pollinating insects. Many crops do not require pollination by insects, but for some, it is either essential or has an influence on yield. The importance of wild bees to pollination is illustrated by Kremen et al. (2021) who demonstrated an increase in almond

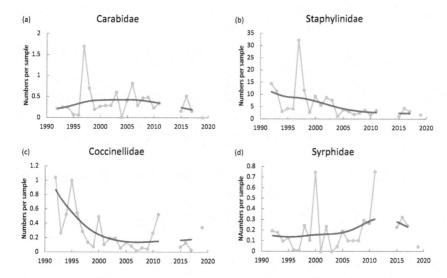

Figure 4.1 Relative abundance of invertebrates in arable crops at Loddington (1992–2019) for (a) ground beetles, (b) rove beetles, (c) ladybirds and (d) hoverflies. No data for 2012, 2013, 2014 or 2018 (Moreby, Ewald et al., n.d.)

yield in response to increasing wild bee numbers, even in the presence of honeybees. Of more direct relevance to temperate cropping systems, Pywell et al. (2015) demonstrated that the yield of field beans was influenced by levels of pollination, Perrot et al. (2018, 2019) demonstrated impacts on yield of oilseed rape and sunflower, and Bullock et al. (2021) also found a positive effect on oilseed rape yield. The diversity of pollinator species can be as important as pollinator abundance and Brittain et al. (2013) and Woodcock et al. (2019) illustrate how this functional diversity can increase the yield of the crops being pollinated.

Storkey et al. (2021) suggest that wild bee species diversity in the UK had declined before any monitoring began in the 1970s and that the wild bee community on farmland today is dominated by generalist species more than was previously the case. There is therefore a need for an understanding of wild bee ecology to inform conservation of both the abundance and diversity of species that influence crop production and conservation of rarer species for their intrinsic biodiversity value (Scheper, 2015).

Surveys at Loddington reveal how different species use a wide range of plant species as a foraging resource through the summer (Cadoux & Szczur, 2016). For example, wild bees forage from dandelion *Taraxacum vulgare* in field margins early in the season, followed by white clover *Trifolium repens* and oilseed rape, field beans and tufted vetch *Vicia cracca* as these become available, and trees such as maples *Acer* spp. and oaks *Quercus* spp. A range of flowering species ensures that a foraging resource is available to many bee species throughout their active period, and to other invertebrates such as butterflies and moths.

In May and June 2016, we trapped 32 solitary bees, comprising mainly *Anthophora plumipes* (11 individuals) and *Andrena* species (14 individuals) and identified pollen grains removed from them (Cadoux & Szczur, 2016). The most commonly occurring plant species represented were trees (maple, oak *Quercus* spp. and fruit trees) and arable crops (oilseed rape and field beans) (Figure 4.2). Sixty percent of pollen gathered from *Andrena nigroaena*, *A. nitida* and *A. bucephala* was from tree species. *Anthophora plumipes* was the main species associated with pollen from crop plants. A wide range of wild herbaceous plant species was also represented as being used by the bees. Most bee species had gathered pollen almost exclusively from just one plant species on any one foraging bout, but *Anthophora plumipes* was the exception, with pollen from a range of plant species being recorded on the individuals sampled. These findings represent only a snapshot of the plant species visited by a solitary bee community at Loddington for one period of time. However, they illustrate the range of plant species being utilised by foraging solitary bees and indicate the differences between bee species in the potential for pollination of both crop and non-crop herbaceous and woody plants.

As well as potential benefits to crop production through enhanced pollination, wild bees are also important for fruit set in hedgerow shrubs. Bumblebees, especially the queens emerging from hibernation in the spring, are important pollinators of blackthorn *Prunus spinosa*, solitary bees are

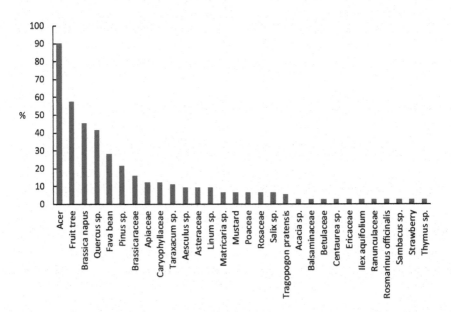

Figure 4.2 Frequency of occurrence of pollen grains found on solitary bees (Szczur & Cadoux, 2016)

pollinators of hawthorn *Crataegus monogyna* later in the year and later still, wasps are major pollinators of ivy *Hedera helix*. Jacobs et al. (2009, 2010) investigated the relative role of pollinating insects in fruit set of hedgerow shrubs by using a fine mesh to experimentally exclude insects from some flowers, and by hand pollinating others (supplementary pollination). For blackthorn, no mature fruits developed when insects were excluded from flowers, and only 4% of flowers formed mature fruit when left open for insects. Hand pollination resulted in the highest fruit production of all (26%), suggesting that blackthorn fruit production was strongly limited by pollinator abundance.

Similar results were obtained for hawthorn (Figure 4.3). Hand-pollinated flowers set more fruit (31%) than those left open (12%), which in turn set more fruit than flowers from which insects were excluded (5%). For ivy, hand pollination did not increase fruit set (56%) significantly over that of flowers left open (45%), but the fruit set of bagged flowers was significantly lower than in the other treatments (3%). This suggests that, although insects are essential for ivy berry production, late summer insect abundance is not limiting fruit set.

For hawthorn, there was not only a clear relationship between the abundance of solitary bees and the abundance of berries in the following autumn but also this correlation was extended to the abundance of migratory thrushes (fieldfares *Turdus pilaris* and redwings *T. iliacus*) in the first half of the winter. There is therefore a direct link between the abundance of wild bees in summer and the abundance of birds in the following winter (Jacobs, 2008). The results also suggest that by influencing the type of berry food available to birds, the insect community present in summer can influence the bird species present in winter, as well as their overall numbers.

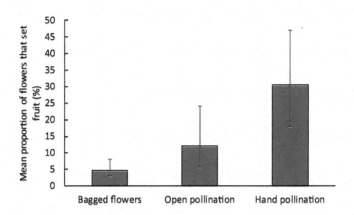

Figure 4.3 Hawthorn fruit set for flowers that were hand pollinated, open (naturally) pollinated, and flowers from which insects were excluded (drawn from data in Jacobs, 2009)

Hedges offer important foraging resources for wild bees and the amount of hedgerow network in an agricultural landscape can influence the level of crop pollination (Kremen et al., 2021). However, hedges are usually dominated by a small number of shrub species resulting in mass flowering for limited periods of time and their value as a foraging resource will depend on the synchronisation with herbaceous flowering plants in the landscape, and whether hedges facilitate or reduce crop pollination will depend on flowering synchrony of hedges and crops (Kremen et al., 2021). Hedges also offer nesting resources for many species and the proportion of semi-natural habitats such as hedges and botanically diverse grassland has been shown to increase wild bee species richness, especially where bee numbers are high (Gaba et al., 2021). Bartual et al. (2019), working across a wide range of European sites, identified the importance of semi-natural habitats to the abundance and species richness of wild bees and other beneficial invertebrates and highlighted the need to recognise that these habitats differ in the extent to which they influence the abundance of pollinating insects.

In 2020, we carried out a comprehensive landscape scale survey of wild bees and hoverflies comprising 56 100-m transects across an area of approximately 30-km^2 in the Eye Brook catchment (Szczur et al., 2021). Bee abundance varied considerably between sites, both at Loddington and across the wider landscape (Figure 4.4). There were substantial differences in numbers of bees recorded in early summer and late summer transects and these differences were not consistent between sites, but sites at which the highest numbers of bees were recorded tended to be those with the most complex semi-natural habitats. The highest bee numbers were associated with abundant foraging resources which included a flowering field bean crop and flowering thistles *Cirsium* spp. along a ditch, or a combination of high-quality foraging habitats in close proximity in the form of a south-facing woodland

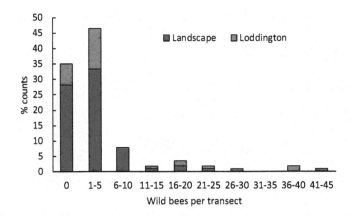

Figure 4.4 Wild bees recorded per transect at Loddington and in the surrounding landscape of the Eye Brook catchment (from Szczur et al., 2021)

edge, a young plantation and a wildlife cover crop including *Phacelia*, buck-wheat and vetches. There was a significant correlation between wild bee abundance and species richness, with those sites supporting the highest bee numbers also being associated with the largest number of species. The rarer species recorded included the nationally scarce bee *Andrena tibialis* and two locally scarce species, *Andrena labialis* and *Chelostoma campanularum.*

Hoverflies were also recorded along the same transects at the same time and revealed the presence of rare species including *Orthonevra nobilis* (long-horned orthonevra), *Platycheirus aurolateralis* and *Volucella inflata* (orange-belted plumehorn), each at just one site. Across all the transects, there was no relationship between the abundance of wild bees and that of hoverflies, but hoverfly species richness was highly correlated with wild bee species richness, implying that landscape scale habitat management targeted at wild bees would also benefit hoverfly conservation.

We used the Integrated Valuation of Ecosystem Services and Tradeoffs (InVEST) pollinator model (Sharp et al., 2020; based on Lonsdorf et al., 2009) to explore the relationship between land cover and the wild bee community (Rayner et al., 2021). We parameterised the model for four guilds of wild bees: (i) ground-nesting bumblebees, (ii) tree-nesting bumblebees, (iii) ground-nesting solitary bees and (iv) cavity nesting solitary bees, following the criteria of Gardner et al. (2020). This approach uses index values for the floral and nesting resources for each land-use class, as well as the nesting preferences and forage distances for each guild, based on expert opinion. We then used the transect data from Szczur et al. (2021) to validate the model.

The model was able to predict the abundance of ground-nesting bumble-bees, tree-nesting bumblebees, and ground-nesting solitary bees. It did not successfully predict the abundance of cavity nesting solitary bees, possibly because this was the least well-represented group recorded on the transects. Tree-nesting bumblebee abundance was most influenced by woodland cover, as expected as, in the UK, this category is represented by just one specialist tree-nesting species. Overall, field boundary habitats had the largest effect on the index of wild bee abundance, providing both a foraging resource and a stable nesting habitat for many species. Grassland had the least effect on abundance, contrasting with studies from some other areas, but probably reflecting the low plant species diversity and limited foraging resource represented by most grass fields in the local landscape. While the sensitivity analysis of the model shows the importance of habitats in isolation, the spatial mapping output from the model also demonstrates how the configuration of land-covers, combined with their area, influences wild bee abundance.

In addition to these diurnal pollinators, Walton et al. (2020) highlight the role moths play in pollination of a wide range of wild plants, in addition to their intrinsic conservation value. Moths have been monitored at Lodding-ton since 1995 using a Rothamsted light trap as part of the national monitoring network. This network of traps reveals a 33% decline in the abundance of macro-moths over the 1968–2017 period (Butterfly Conservation, 2021). Forty-one percent of species had declined over this period, while 10% had increased, the remainder showing no statistically significant trend. In contrast, data from the trap at Loddington show that the overall abundance of macro-moths was 36% higher in 2019 than it was in 1995. The number of species represented in catches has also increased over this period, with an average of 173 for the most recent 12-year period, compared to an average of 146 for the first 12 years of monitoring.

Two species which illustrate this trend are the flounced rustic *Luperina testacea* which has declined nationally by 60% since 1969, and the brown rustic *Rusina ferruginea* which has declined by 64% nationally over the same period. Both species have increased substantially at Loddington since 2007, although these increases were checked in 2012 by prolonged summer rainfall before increasing again in subsequent years (Figure 4.5). Flounced rustic is associated with dry grassland and may have benefited from the extensive creation of wildlife habitat at Loddington such as wild flower margins, beetle banks and grass margins, including a 12 metre margin close to the moth trap. A significant increase in another kind of grassland species, the straw dot *Rivula sericealis,* can also be attributed to the increase in this habitat. None were caught in the first five years, 87 in the second five years and 986 in the third five years of monitoring. Brown rustic is associated with herbaceous vegetation in woodland and is likely to have benefited from the planting of new woods and the thinning of existing woodland to encourage the herb and shrub layer for game and other wildlife (Szczur, 2021).

Oak *Quercus robur* has featured prominently in the woodland planting over the past 30 years and is a tree species that was previously scarce at

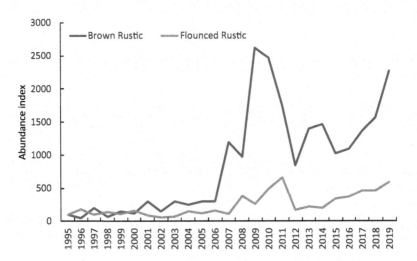

Figure 4.5 Changes in the relative abundance of brown rustic and flounced rustic, 1995–2019

Loddington. This is the likely explanation for occasional recent records of merveille du jour *Dichonia aprilina*, black arches *Lymantria monacha* and oak nycteoline *Nycteola revayana*. Other species that have been recorded only in recent years and which are associated with individual tree species that have been planted on the farm include scorched carpet *Ligdia adustata* (spindle, *Euonymus europaeus*), white satin *Leucoma salicis* and chocolate-tip *Clostera curtula* (poplars *Populus* spp. and willows *Salix* spp.). Climate change has also contributed to the range expansion of the latter. Hedge bedstraw *Galium mollugo* and lady's bedstraw *Galium verum* have been included in flower-rich margins and may explain the substantial increase in green carpet moth *Colostygia pectinataria*, for which bedstraws are the larval food plant. Flame shoulder *Ochropleura plecta* also exploits bedstraws, although not exclusively, and shows a similar trend (Szczur, 2021).

Other habitat changes that are likely to have influenced moth abundance include the planting of new hedges and the increase in size of many existing ones. Working elsewhere in the UK, Merckx et al. (2009a, 2009b) reported twice as many moths in field margins as in field centres, and positive effects of hedgerow trees on both moth abundance and diversity. In addition, the abundance of grass, herb and lichen feeding moths was significantly greater in 6-m wide grass margins than those that were 1 m wide.

In field margins with trees, Merckx et al. (2010) reported a greater abundance of moth species whose host plants were trees, shrubs or climbers (mostly Geometridae) than in field margins without trees, but there was no such difference for species (mainly Noctuidae) whose host plants were grasses and herbs. This is likely to result from an increased shelter as much as the availability of food plants as Geometridae are less mobile than Noctuidae. Coulthard et al. (2016) observed moths along transects that were 1, 5 and 10 m parallel from hedges and reported that 68% of moths recorded were adjacent to the hedge, and of those 69% were moving parallel to it. They also noted that gaps in hedges were important to facilitate moth movement between fields.

Like moths, butterflies are not amongst the invertebrates that are most strongly associated with the land used to produce food but are an integral component of farmland ecosystems, mainly using the network of interstitial habitats between fields. From 1997 to 2001, butterfly transects were carried out at Loddington at weekly intervals between April and September along arable field margins, managed set-aside, new woodland, and a railway embankment. Arable margins were the least used habitat but set-aside strips that were managed for wildlife supported the highest numbers of both individuals and species (Bence, 1997–2001). While highlighting the potential for butterflies of habitat creation on arable land, shelter, nectar sources and larval food plants associated with perennial vegetation are also known to be major determinants of butterfly abundance and species richness on farmland (Dover, 2021; Pywell et al., 2004).

Invertebrates as bird food

Invertebrates present in the crops themselves are monitored annually using suction sampling, revealing a range of species according to the crop types and their management. Many of these invertebrates provide food for birds using farmland habitats. Although many farmland birds feed on seeds for most of the year, most are dependent on insect food during the breeding season. This has been most convincingly demonstrated by intensive long-term GWCT research into grey partridge *Perdix perdix* ecology which has been carried out since the 1960s and helped to inform subsequent research at Loddington and elsewhere (Potts, 1986, 2012).

Research carried out in Sussex and Hampshire in southern England showed that grey partridge chick growth and survival rates in the first two weeks of life were highly dependent on a diet comprising almost exclusively invertebrates. The research and subsequent long-term monitoring in Sussex have also shown that there has been a substantial decline in the abundance of invertebrates in arable crops associated, not just with the use of broad-spectrum insecticides, but with herbicides which remove essential invertebrate food plants in crops (Potts et al., 2015, 2020).

Key to the research on grey partridge chick diet is a method of faecal analysis in which faeces are collected from partridge brood roost sites and analysed under the microscope. Based on a comprehensive knowledge of invertebrate hard parts it is possible to build up a detailed picture of the invertebrates that contribute to chick diet. At Loddington, we also used this method in the 1990s to determine the diet of wild pheasant *Phasianus colchicus* chicks from 15 broods on the farm (Bence & Moreby, n.d.). We also extended the approach to include songbird species, yellowhammer *Emberiza citrinella* (67 broods), dunnock *Prunella modularis* (23 broods) and whitethroat *Sylvia communis* (36 broods), collecting faecal samples from nestlings when nests were visited as part of our routine monitoring (Moreby & Stoate, 2000, 2001). As a result, we can compare the diet of these four species at the same site (Figure 4.6).

Beetles form the main component of the diet for pheasant, yellowhammer and dunnock but rather less so for whitethroat. Curculonidae (weevils) formed the bulk of these for pheasant and dunnock, while adult and larval Chrysomelid (leaf) beetles were also important to pheasant. Carabid and Staphylinid beetles were more important to yellowhammers which also fed on Tipulids (craneflies) more than the other bird species did. Aphids and Heteroptera (plant bugs) featured prominently in pheasant chick diet, and the diet of pheasant chicks and dunnock also included Symphyta (sawfly) larvae. Butterfly and moth caterpillars featured in the diet of all species and formed the largest component for whitethroats. Spiders were represented mainly in the diet of yellowhammer and whitethroat.

Caterpillars associated with the increasing numbers of moths across the farm are likely to have benefited those bird species such as whitethroat

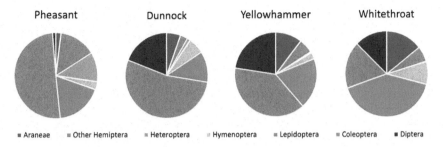

Figure 4.6 Relative abundance of various invertebrate taxa in the diet of pheasant chicks and nestlings of yellowhammer, dunnock and whitethroat

which forage in hedges and associated grass margins. Our long-term monitoring of invertebrates in arable fields reveals a fluctuating abundance of both Lepidoptera and Symphyta larvae (Figure 4.7). While their abundance at low density in crops will benefit those bird species using cropped land as foraging habitat, there does not appear to have been an increase in this food availability for provisioning birds over the 30 years of the Allerton Project.

In 2002 and 2003, we found Coleoptera forming nearly 70% of the tree sparrow *Passer montanus* nestling diet in both years, of which about half comprised Carabid beetles (Toomey, 2003). The faecal analysis revealed that Diptera, mainly Tipulidae, represented the next largest component.

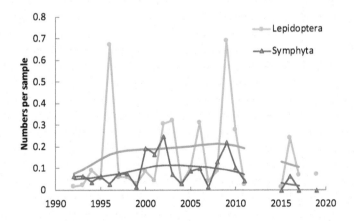

Figure 4.7 Relative abundance of Lepidoptera and Symphyta larvae in arable crops, 1992–2019. No data for 2012, 2013, 2014 and 2019 (Moreby, Ewald et al., n.d.)

We used a different method to look at the nestling diet of skylarks *Alauda arvensis*, in this case using neck ligatures under licence for short periods to sample the invertebrates as they were fed to nestlings (Murray, 2004). Thirty-one percent of the skylark invertebrate diet comprised insect larvae which were fed to 75% of the 31 broods monitored between 2000 and 2002. The next most important groups were adult Coleoptera, forming 18% of all samples, and spiders constituting 14% of samples. The latter were used more than expected from their availability within the foraging areas. Adult Diptera comprised 9% of the diet, as with tree sparrows, mainly in the form of Tipulids.

Breeding birds

Breeding pheasants were studied intensively in the 1990s when there was an established wild population. Radio-tracking of 75 hens and 32 broods was used to find out about nest sites, nest survival rates, and the fate of broods of chicks. This work revealed that chick survival and brood size were both negatively correlated with pheasant hen density, the implication being that the supply of invertebrate food was limiting chick survival and that competition led some pheasants to use areas with low invertebrate food abundance when hen numbers were higher (Figure 4.8). This density dependence principle may also apply to other bird species, with chick survival being highest when breeding densities are lower because only the 'best' foraging areas of the farm are occupied. However, this is less likely to apply where nest site characteristics, rather than foraging habitats, are the main determinant of breeding site selection.

For yellowhammers, for example, the location of breeding territories on farmland at Loddington is largely determined by the presence of ditches

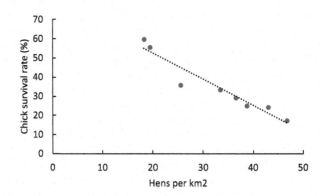

Figure 4.8 Pheasant chick survival rate in relation to breeding hen pheasant density (Anon, 2001)

with abundant herbaceous vegetation in field boundaries (Stoate & Szczur, 2001), even though most of the foraging is carried out in adjacent fields. Yellowhammers at Loddington in the 1990s changed their use of foraging habitat through the breeding season, gathering invertebrate food first in oilseed rape, then in barley and finally in wheat (Stoate et al., 1998). As well as invertebrates, yellowhammers feed unripe cereals to nestlings, and this might explain the switch into barley and wheat as these crops ripen in June and July, although insect abundance in these crops will remain important. We know from our observations that the maximum distance travelled by provisioning yellowhammers is 300 metres, so any single yellowhammer territory therefore enjoys a longer effective foraging period where several crops are present within that range.

Observations of 16 tree sparrow nests in 2002 and 2003 revealed that birds were foraging in commercial crops more than expected from their availability, travelling up to 220 metres from the nest, even though other habitats such as wild bird seed crops were closer to the nest sites (Toomey, 2003).

Nesting skylarks at Loddington have a foraging range of up to about 230 metres and analysis of foraging observation data from 16 nests in 1999 revealed that linseed and set-aside (fallow) land were used more than expected from their availability, while winter wheat was used less (Bence, 1999). More detailed data collected over the next two years revealed that the structure of the vegetation influenced where skylarks foraged, with relatively open structured vegetation and bare ground being favoured, more than was the case for yellowhammers (Murray, 2004; Murray et al., 2002). This is in line with other studies identifying that foraging habitat structure is at least as important as invertebrate food abundance for many bird species (Wilson et al., 2005). In our study, the selection of foraging areas by skylarks was influenced by the abundance of invertebrates, especially spiders, but this was not the case for yellowhammers. Together with their greater use of cereal crops, this suggests that yellowhammers were favouring areas where they could gather both invertebrates and unripe grain, whereas grain was not important for skylarks.

Although skylarks seemed to be influenced by spider abundance in the selection of foraging areas, and spiders were sometimes a major component of nestling diet, the proportion of spiders in the diet was related negatively to the growth rates of nestlings. Spiders may be sub-optimal in terms of the nutritional requirements of nestlings, and their ready availability, or the vegetation structure they are associated with, may have been a greater influence on their selection by adults than their nutritional value.

Invertebrates known to be favoured by farmland birds in other parts of the UK and continental Europe, such as sawfly larvae and grasshoppers, probably formed only a very small part of nestling diet at Loddington simply because they were not available in sufficient numbers at the time. In other parts of Europe where caterpillars, grasshoppers and crickets are more abundant, they feature prominently in the diet of farmland birds and influence their habitat use (e.g. Stoate et al., 2000).

We studied habitat use by breeding song thrushes *Turdus philomelos* by using radio-tracking to overcome the secretive nature of the species for which direct observation of foraging behaviour is unrealistic (Murray, 2004). Although occupying farmland, song thrushes favoured boundary habitats and woodlands as nest sites and foraging areas. Male thrushes had larger foraging ranges than females. The area of pasture within the foraging range had a positive influence on nest survival, and the area of cereals had a negative influence, especially during incubation (Figure 4.9). Pasture was used as a foraging habitat, but not any more than would be expected from its availability. Boundary habitats and woodland were more strongly used and there was some indication that thrushes were selecting sites with higher earthworm abundance. Pasture may dry out faster than woodland and field boundary habitats, making earthworms less available, and the greater shrubby cover in woodland and hedges also offers protection from avian predators that is not available in pasture. Certainly, the cereal area and other arable land supported lower earthworm densities than pasture, woodland and field boundaries at the time of the study, as discussed in Chapter 3.

For other bird species, cereal crops are often used less as a foraging habitat than break crops within the same rotation. Break crops such as oilseed rape and field beans are used as a foraging habitat by many bird species because of the invertebrates that they support. In a comparison of oilseed rape, field beans and hemp fields, at Loddington and another local farm, bird species consistently recorded foraging in break crops comprised blackbird *Turdus merula*, dunnock, yellow wagtail *Motacilla flava*, whitethroat, linnet *Linaria cannabina*, chaffinch, tree sparrow, reed bunting *Emberiza schoeniclus* and yellowhammer (Stoate et al., 2012). While these species are

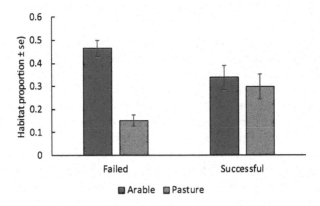

Figure 4.9 Areas of arable and pasture within the foraging range of successful and failed song thrush nests (drawn from data in Murray, 2004)

predominantly foraging for invertebrates, linnets represent an exception in that they are gathering unripe rape seeds to feed nestlings. Hemp supported the lowest range of bird species, with whitethroat being the dominant species, probably because the dense vertical structure of the mature crop makes it inaccessible for other species. However, at both study sites in late summer, this crop supported pre-migratory swallow roosts of more than 1,000 birds.

Although some of the bird species present on the farm in summer are migratory, most are resident on the farm in winter. The majority of these are feeding on seeds. The seasonal pattern of abundance for each species varies from year to year as birds are mobile and numbers fluctuate in response to weather conditions and food availability. In the case of migratory birds, weather conditions elsewhere in the wintering range can influence numbers of birds present in the UK. We have demonstrated in this chapter that the abundance of pollinating insects in summer can influence numbers and distribution of fruit-eating birds in winter, but seed-eating bird numbers on farmland are determined by the availability of seeds associated with arable weeds, spilt grain or perennial plants outside the cropped area. This is discussed in more detail in Chapter 5.

Mammals

Although most research and monitoring of vertebrates at Loddington has focussed on birds, several mammal species are also present. Brown hares *Lepus europaeus* not only make regular use of commonly occurring crops such as winter wheat and oilseed rape but also benefit from crop diversity, especially where grass is present in arable landscapes, as suitable food is present throughout the year. Day and night-time radio-tracking of 11 hares at Loddington in 1997 revealed that set-aside was the most used habitat relative to its availability, with winter beans being the least used (Tapper, 1997).

Muntjac *Muntiacus muntjac* is the only deer species occurring at and around Loddington having first colonised the area in the 1970s. Muntjac are regularly seen on farmland but are mainly associated with areas of woodland. Stoate et al. (2005) surveyed the damage to bluebells *Hyacinthoides non-scripta* in the larger woods and found that, while this varied between woods, the main influence was proximity to shrub cover. The woods are thinned periodically to allow regeneration of the herb layer, including bluebell and rarer plants for which they are designated Sites of Special Scientific Interest (SSSI), but the process of succession results in increased shrub cover for muntjac. The proportion of damaged plants was also highest at low plant densities which may imply that rarer species are more susceptible to damage.

Wood mouse *Apodemus sylvaticus*, harvest mouse *Micromys minutus*, bank vole *Myodes glareolus*, field vole *Microtus agrestis*, common shrew *Sorex araneus*, pygmy shrew *S. minutus* and water shrew *Neomys fodiens* are all present on farmland at Loddington, with live trapping of 1,119 individuals in September 1998 providing a means of estimating densities at the time of 84 per hectare for wood mouse, 35 per hectare for bank vole, 26 per hectare for field vole, 65 per hectare for common shrew, and 13 per hectare for water shrew (Bence, 1998). The numbers of pygmy shrew were too low to estimate densities. All species except harvest mouse were widely distributed across habitats, with field vole and common shrew numbers being influenced by the amount of vegetative cover at the trap site. In an additional survey of harvest mouse nests, Bence et al. (2003) found a nest density of 14.3/ha of field margin habitat, with nests normally being built where shrubby stems provided additional structural support to the surrounding grasses.

We used passive bat detectors in 2021 to provide a quantitative estimate of relative abundance of bats and understand better their landscape scale use of habitats (Cossa, 2021). Common pipistrelle *Pipistrellus pipistrellus* is by far the most common bat species, representing 62% of the 178,000 records, with soprano pipistrelle *P. pygmaeus* also being common (23%). Unidentified *Myotis* bats are regularly recorded (5.82%), as are noctules *Nyctalus noctule* (3.66%) and brown long-eared bats *Plecotus auritus* (1.34%). Barbastelle *Barbastella barbastellus* (0.35%) has established roost sites in the larger woods and Leisler's bat *N. leisleri* (0.23%) and serotine *Eptesicus serotinus* (0.01%) have also been recorded. At least eight species are therefore known to occur locally, with Leisler's and barbastelle representing nationally scarce species.

Bats are predominantly associated with woodland, especially two ancient semi-natural SSSI woods close to the northern part of the farm, but farmland is also used as a foraging habitat and as corridors for movement between woods. The peak activity along farmland hedges is in the early part of the night when bats are moving out from roost sites (Figure 4.10), but the largest hedges and shelter belts are also used for foraging.

Working in southern England, McHugh et al. (2018) have shown that pipistrelle bats make use of field boundaries as foraging habitats as well as commuting routes across agricultural landscapes, with differences in species response to the presence of agri-environment schemes habitats such as wild flower, pollen and nectar, or grass margins. The use of these woody and herbaceous habitats on farmland by species that are primarily associated with woodland illustrates the importance of habitat configuration as well as diversity. This is in line with the findings for wild bees referred to earlier.

Our work in the Alentejo region of southern Portugal (Santana et al., 2017) identified the area of low input productive land, rather than habitat diversity or configuration, as being the main influence on the abundance of bird species. This can be explained by the presence of species such as great bustard *Otis tarda*, little bustard *Tetrax tetrax*, calandra lark *Melanocorypha calandra* and corn bunting *Emberiza calandra* which are strongly dependent on open steppe landscapes and the low intensity of crop management. However, even here, there was some influence of landscape configuration on overall bird species richness. The greater influence of habitat

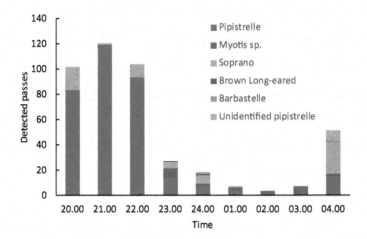

Figure 4.10 Example of the nocturnal use of a hedgerow by bat species present at Loddington (based on passive bat detector records) (Cossa, 2021)

configuration for mammals, birds and invertebrates in areas such as lowland England can be explained by more intensive crop management and the presence of species that are distributed along a spectrum from farmland specialists to species associated mainly with woodland, and by the fact that there are short-term temporal and longer-term seasonal differences in habitat use by these species, as illustrated in this chapter.

Notes

1 This issue is discussed in Chapter 7.
2 This issue is discussed further in Chapter 8.
3 *Pterostichus madidus* and *Nebria brevicollis* were the most abundant species in woodland edges, while *Patrobus atrorufus* and *Harpalus rufipes* were the most abundant species in hedges. *Anchomenus dorsalis, Abax parallelepipedus, Bembidion obtusm, Ophonus puncticeps* and *Synuchus vivalis* were only found in the woodland edges, and *Amara similata* was only found in hedges. 96%, 77% and 75% of *Trechus quadristriatus, Harpalus rufipes* and *Carabus violaceus*, respectively, were found in hedges, and 95%, 88% and 86% of *Pterostichus madidus, Leistus spinibarbis* and *Nebria salina* caught in pitfall traps were found in woodland edge.

References

Anon (2001) The Allerton project. *The Game Conservancy Trust Review of 2000* 132, 93–102.
Bartual, A. M., Sutter, L., Bocci, G., Moonen, A. C., Cresswell, J., Entling, M. H., Giffard, B., Jacot, K., Jeanneret, P., Holland, J. M., Pfister, S. C., Pintér, O., Veromann, E., Winkler, K., Giffard, B. & Albrecht, M. (2019) The potential of

different types of semi-natural habitats types to sustain pollinators and natural enemies in European agricultural landscapes. *Agriculture, Ecosystems & Environment* 279, 43–52.

Bence, S. (1997–2001) Unpublished data.

Bence, S. (1998) Unpublished data.

Bence, S. (1999) Unpublished data.

Bence, S. & Moreby, S. (n.d.) Unpublished data.

Britain, C., Bremen, C. & Klein, A. M. (2013) Biodiversity buffers pollination from changes in environmental conditions. *Global Change Biology* 19, 540–547.

Brown, P. J. & Roy, H. E. (2018) Native ladybird decline caused by the invasive harlequin ladybird *Harmonia axyridis*: evidence from a long-term study. *Insect Conservation and Diversity* 11 (3), 230–239.

Bullock, J. M., McCracken, M. E., Bowes, M. J., Chapman, R. E., Graves, A. R., Hinsley, S. A., Hutchins, M. G., Nowakowski, M., Nicholls, E., Oakley, S., Old, G. H., Ostle, N. J., Redhead, J. W., Woodcock, B. A., Bedwell, T., Mayes, S., Robinson, V. S. & Pywell, R. F. (2021) Does agri-environmental management enhance biodiversity and multiple ecosystem services?: a farm-scale experiment. *Agriculture, Ecosystems and Environment* 320, 107582. DOI:10.1016/j.agee.2021.107582

Cadoux, N. & Szczur, J. (2016) Unpublished data.

Collins, K. L., Boatman, N. D., Wilcox, A., Holland, J. M. & Chaney, K. (2002) Influence of beetle banks on cereal aphid predation in winter wheat. *Agriculture, Ecosystems and Environment* 93, 337–350.

Collins, K. L., Boatman, N. D., Wilcox, A. & Holland, J. M. (2003) A five-year comparison of overwintering polyphagous predator densities within a beetle bank and two conventional hedge banks. *Annals of Applied Biology* 143, 63–71.

Comont, D., Lowe, C., Hull, R., Crook, H. L., Onkokesung, N., Beffa, R., Childs, D. Z., Edwards, R., Freckleton, R. P. & Neve, P. (2020) Evolution of generalist resistance to herbicide mixtures reveals a trade-off in resistance management. *Nature Communications* 11, 3086. DOI:10.1038/s41467-020-16896-0

Cossa, N. (2021) Unpublished data

Dover, J. W. (2021) The Ecology of Butterflies and Moths in Hedgerows and Field Margins. In: Dover, J. W. (Ed.) *The Ecology of Hedgerows and Field Margins*. Routledge, Abingdon. 187–209.

Ewald, J. A., Aebischer, N. J., Moreby, S. J., Duffield, S. J., Crick, H. Q. P. & Morecroft, M. B. (2015) Influences of extreme weather, climate and pesticide use on invertebrates in cereal fields over 42 years. *Global Change Biology* 21, 3931–3950.

Ewald, J. A., Wheatley, C. J., Aebischer, N. J., Duffield, S. J. & Heaver, D. (2016). *Investigation of the Impact of Changes in Pesticide Use on Invertebrate Populations*. Natural England Commissioned Report, NECR182. Natural England, York.

Ewald, J. A., Aebischer, N. J. & Sotherton, N. (2021) Invertebrate trends in an arable environment: long-term changes from the Sussex Study in Southern England. In: Hurford, C., Wilson, P. & Storkey, J. (Eds.) *The Changing Status of Arable Habitats in Europe*. Springer Nature, Cham. 157–172. http://dx.doi.org/10.1007/978-3-030-59875-4_11

Finch, G. (2017) Unpublished data.

Finch, D., Schofield, H., Floate, K. D., Kubasiewicz, L. M. & Mathews, F. (2020) Implications of endectocide residues on the survival of Aphodiine dung beetles: A meta-analysis. *Environmental Toxicology and Chemistry* 39, 863–872.

Gaba, S. & Bretagnolle, V. (2021) Designing Multifunctional and Resilient Agricultural Landscapes: Lessons from Long-Term Monitoring of Biodiversity and Land Use. In: Hurford, C., Wilson, P. & Storkey, J. (Eds.) *The Changing Status of Arable Habitats in Europe.* Springer Nature, Cham. 203–224.

Gardner, E., Breeze, T. D., Clough, Y., Smith, H. G., Baldock, K. C., Campbell, A., Garratt, M. P., Gillespie, M. A., Kunin, W. E. and McKerchar, M. (2020) Reliably predicting pollinator abundance: challenges of calibrating process-based ecological models. *Methods in Ecology and Evolution* 11, 1673–1689. DOI:10.1111/2041-210X.13483

Haughton, A. J. (2000) *Investigations into spray drift and the effects of herbicide on non-target arable field margin arthropods.* Unpublished PhD thesis. Open University.

Holland, J. (2019) Unpublished data.

Holland, J. (2021) Unpublished data.

Jacobs, J. H. (2008) *The birds and the bees: pollination of fruit-bearing hedgerow plants and consequences for birds.* Unpublished PhD thesis. University of Stirling.

Jacobs, J. H., Clark, S., Denholm, I., Goulson, D., Stoate, C. & Osborne, J. (2009) Pollination of fruit-bearing hedgerow plants and the role of flower-visiting insects in fruit set. *Annals of Botany* 104 (7), 1397–1404.

Jacobs, J. H., Clark, S., Denholm, I., Goulson, D., Stoate, C. & Osborne, J. (2010) Pollination and fruit set in common ivy Hedera helix (Araliaceae). *Arthropod-Plant Interactions* 4, 19–28.

Kremen, C., Albrecht, M. & Ponisio, L. (2021) Restoring pollinator communities and pollination services in hedgerows in intensively managed agricultural landscapes. In Dover, J. D. (Ed.) *The Ecology of Hedgerows and Field Margins.* Routledge, Abingdon. 163–185.

Lonsdorf, E., Kremen, C., Ricketts, T., Winfree, R., Williams, N. & Greenleaf, S. (2009) Modelling pollination services across agricultural landscapes. *Annals of Botany* 103, 1589–1600.

Manning, P., Slad, E. M., Benyon, S. A. & Lewis, O. T. (2017) Effect of dung beetle species richness and chemical perturbation on multiple ecosystem functions. *Ecological Entomology* 42, 577–586.

McHugh, N., Bown, B. L., Forbes, A. S., Hemsley, J. A. & Holland, J. M. (2018) Use of agri-environment scheme habitats by pipistrelle bats on arable farmland. *Aspects of Applied Biology* 139, 15–22.

Merckx, T., Feber, R. E., Dulieu, R. L., et al. (2009a) Effect of field margins on moths depends on species mobility: field-based evidence for landscape-scale conservation. *Agriculture, Ecosystems & Environment* 129, 302–309.

Merckx, T., Feber, R. E., Riordan, P., et al. (2009b) Optimizing the biodiversity gain from agri-environment schemes. *Agriculture, Ecosystems and Environment* 130, 177–182.

Merckx, T., Feber, R. E., McLaughlan, C., et al.(2010) Shelter benefits less mobile moth species: the field-scale effect of hedgerow trees. *Agriculture, Ecosystems & Environment* 138, 147–151.

Moreby, S. J. & Stoate, C. (2000) A quantitative comparison of neck collar and faecal analysis to determine passerine nestling diet. *Bird Study* 47, 320–331.

Moreby, S. J. & Stoate, C. (2001) Relative abundance of invertebrate taxa in the nestling diet of three Farmland Passerine species, Dunnock *Prunella modularis*,

Whitethroat *Sylvia communis* and Yellowhammer *Emberiza citrinella*. *Agriculture, Ecosystems & Environment* 86, 125–134.

Moreby, S., Ewald, J., et al. (n.d.) Unpublished data.

Murray, K. A., Wilcox, A. & Stoate, C. (2002) A simultaneous assessment of skylark and yellowhammer habitat use on farmland. *Aspects of Applied Biology* 67, 121–127.

Murray, K. A. (2004) *Factors affecting foraging by breeding farmland birds*. Unpublished PhD thesis. Open University.

Perrot, T., Gaba, S., Roncoroni, M., Gautier, J. L. & Bretagnolle, V. (2018) Bees increase oilseed rape yield under real field conditions. *Agriculture, Ecosystems & Environment* 266, 39–48.

Perrot, T., Gaba, S., Roncoroni, M., Gautier, J. L., Saintilan, A. & Bretagnolle, V. (2019) Experimental quantification of insect pollination on sunflower yield, reconciling plant and field scale estimates. *Basic Applied Ecology* 34, 75–84.

Potts, G. R. (1986) *The Partridge: Pesticides, Predation and Conservation*. Collins, London.

Potts, G. R. (2012) *Partridge: Countryside Barometer*. Collins, London.

Pywell, R. F., Warman, E. A., Sparks, T. H., et al. (2004) Assessing habitat quality for butterflies on arable farmland in England. *Biological Conservation* 118, 313–325.

Pywell, R. F., Heard, M. S., Woodcock, B. A., Hinsley, S., Ridding, L., Nowakowski, M. & Bullock, J. M. (2015) Wildlife-friendly farming increases crop yield: evidence for ecological intensification. *Proceedings of the Royal Society B: Biological Sciences* 282 (1816), 8.

Rayner, M., Balzter, H., Jones, L., Whelan, M. & Stoate, C. (2021) Effects of improved land-cover mapping on predicted ecosystem service outcomes in a lowland river catchment. *Ecological Indicators* 133, 108463. DOI:10.1016/j.ecolind.2021.108463

Santana, J., Reino, L., Stoate, C., Moreira, F., Ribeiro, P. F., Santos, J. L., Rotenberry, J. T. & Beja, P. (2017) Combined effects of landscape composition and heterogeneity on farmland avian diversity. *Ecology and Evolution* 7, 1212–1223.

Scheper, J. A. (2015) *Promoting Wild Bees in European Agricultural Landscapes*. Alterra Scientific Contributions 47. Alterra, Wageningen.

Sharp, R., Douglass, J., Wolny, S., Arkema, K., Bernhardt, J., Bierbower, W., Chaumont, N., Denu, D., Fisher, D., Glowinski, K., Griffin, R., Guannel, G., Guerry, A., Johnson, J., Hamel, P., Kennedy, C., Kim, C. K., Lacayo, M., Lonsdorf, E., Mandle, L., Rogers, L., Silver, J., Toft, J., Verutes, G., Vogl, A. L., Wood, S. & Wyatt, K. (2020) InVEST 3.9.0 User's Guide: The Natural Capital Project, Stanford University, University of Minnesota, The Nature Conservancy and World Wildlife Fund.

Stoate, C., Moreby, S. J. & Szczur, J. (1998) Breeding ecology of farmland Yellowhammers *Emberiza citrinella*. *Bird Study* 45, 109–121.

Stoate, C., Borralho, R. J. & Araújo, M. (2000) Factors affecting corn bunting *Miliaria calandra* abundance in a Portuguese agricultural landscape. *Agriculture, Ecosystems & Environment* 77, 219–226.

Stoate, C. & Szczur, J. (2001) Whitethroat *Sylvia communis* and Yellowhammer *Emberiza citrinella* nesting success and breeding distribution in relation to field boundary vegetation. *Bird Study* 48, 229–235.

Stoate, C. (2004) *Lesser marsh grasshoppers at Loddington: a summary report on research by the Allerton Research and Educational Trust and the University of Leicester*. Unpublished report.

Stoate, C., Berry, A. & Lear, A. (2005) *Muntjac damage to ground flora in Leighfield Forest: a report to English Nature and the Forestry Commission.* Unpublished report, GWCT Allerton Project, Loddington.

Stoate, C. (2011) Biogeography of agricultural environments. *The Sage Handbook of Biogeography.* Sage, London. 338–356.

Stoate, C., Szczur, J. & Partridge, J. (2012) The Ecology of Hemp Production Relative to Alternative Break Crops. In: McCracken, K. (Ed.) *Valuing Ecosystems: Policy Economic and Management Interactions.* Scottish Agricultural College & Scottish Environment Protection Agency, Edinburgh. 264–269.

Stoate, C. & Wilson, P. (2021) Historical and Ecological Background to the Arable Habitats of Europe. In: Hurford, C., Wilson, P. & Storkey, J. (Eds.). *The Changing Status of Arable Habitats in Europe.* Springer Nature, Cham. 3–13.

Storkey, J., Brown, M. J. F., Carvell, C., Dicks, L. V. & Senapathi, D. (2021) Wild Pollinators in Arable Habitats: Trends, Threats and Opportunities. In: Hurford, C., Wilson, P. & Storkey, J. (eds.) *The Changing Status of Arable Habitats in Europe.* Springer, Cham. 187–201.

Szczur, J. & Cadoux, N. (2016) Unpublished data.

Szczur, J. (2021) Personal communication.

Tapper, S. (1997) Unpublished data.

Tooley, J. (1999) Unpublished data.

Toomey, M. (2003) *The Breeding Success and Foraging Behaviour of Tree Sparrows (Passer montanus) on an arable farmland site in the East Midlands Region of England.* Unpublished MSc thesis. University of Reading.

Walton, R. E., Sayer, C. D., Bennion, H. & Axmacher, J. C. (2020) Nocturnal pollinators strongly contribute to pollen transport of wild flowers in an agricultural landscape. *Biology Letters* 16, 20190877. DOI:10.1098/rsbl.2019.0877

Willis, C. E., Foster, S. P., Zimmer, C. T., Elia, J., Chang, X., Field, L. M., Williamson, M. S. & Davies, T. G. (2020) Investigating the status of pyrethroid resistance in UK populations of the cabbage stem flea beetle (*Psylliodes chrysocephala*). *Crop Protection* 138, 105316. DOI:10.1016/j.cropro.2020.105316

Wilson, J. D., Whittingham, M. J. & Bradbury, R. B. (2005) The management of crop structure: a general approach to reversing the impacts of agricultural intensification on birds. *Ibis*, 147 (3), 453–463.

Woodcock, B. A., Garratt, M. P. D., Powney, G. D., Shaw, R. F., Osborne, J. L., Soroka, J., Lindstrom, S. A. M., Stanley, D., Ouvard, P., Edwards, M. E., Jauker, F., McCracken, M. E., Zou, Y., Potts, S. G., Rundlof, M., Noriega, J. A., Greenop, A., Smith, H. G., Bommarco, R., van der Werf, W., Stout, J. C., Steffan-Dewenter, I., Morandin, L., Bullock, J. M. & Pywell, R. F. (2019) Meta-analysis reveals that pollinator functional diversity and abundance enhance crop pollination and yield. *Natural Communication* 10, 10.

5 Developing new approaches to wildlife management

Our improved understanding of farmland ecology and the challenges to wildlife conservation and other environmental objectives provides a foundation for the development of appropriate management practices. This has been a central feature of Allerton Project research from the early years. In terms of wildlife, we have been concerned about the provision of ecosystem services such as pollination of crops and wild plants and the natural control of crop pests, as well as the intrinsic cultural value of the wide-ranging biodiversity found on farmland. In doing so, we need to consider the year-round requirements of the species involved, developing the evidence base for ecological benefits of new management practices.

There are also practical considerations, both in terms of the implications of the management practices for farming operations, and in terms of the creation and management of the habitats themselves. For the former, there is a need to reduce negative impacts on productive land use. We have identified that crop yields on the outer 6 metres of the field are generally around 80% of those on the rest of the field (Chaney et al., 1999), while input costs remain the same. From an economic perspective, there is therefore a strong argument for focusing habitat creation on field margins. The majority of farmland species are associated with field margins (Dover, 2019) so in this instance, economic and environmental objectives are well aligned. By concentrating on the less productive parts of the farm, and of individual fields, we can minimise the effect of habitat creation on both overall food production from the farm and economic performance.

Practical aspects of habitat creation and management are important considerations. This is best achieved where the management practiced exploits farmers' existing skillsets and equipment as in the establishment of grass or crops, although inevitably different objectives require slight changes to the methods adopted. The plant species that deliver ecological benefits may be the same or very different from those more normally grown on farms, while the management of them will inevitably be different.

DOI: 10.4324/9781003160137-5

Grass margins and beetle banks

The point where productive land meets semi-natural habitat is normally a strip of perennial grassy vegetation in the field edge. It is common enough now, but when the Allerton Project started in 1992, such vegetation was often absent because it had been sprayed out or destroyed by accidental spray drift. Misplaced fertiliser meant that what vegetation grew in this space was dominated by a small number of species associated with high nutrient conditions such as cleavers *Galium aparine* and sterile brome *Bromus sterilis*, species which were themselves annual arable weeds (Boatman et al., 1994).

Box 5.1 Grass margins

Grass margins can be established by natural regeneration of grasses and other species, or by sowing a specific seed mixture (Figure 5.1). Adopting the former approach on arable land can be risky as the plant community tends to be dominated by annuals and invasive perennials such as docks and thistles which support considerably lower densities of beneficial invertebrates. Sowing a known seed mixture enables a

Figure 5.1 A 6-metre grass margin serving as a buffer strip at the base of an arable slope

(Continued)

more managed approach to be taken, controlling the combination of species present in the sward.

If the primary objective of the margins is to reduce runoff to adjacent water courses, then the vegetation cover should be dense at ground level, although a more open structure is desirable if the margin is intended as a habitat for pollinating insects or for breeding birds that are gathering invertebrate food for dependent young. Grass species such as cocksfoot *Dactylis glomerata*, meadow fescue *Festuca pratensis* and tall fescue *F. arundinacea* fulfil the former role as these are dense or tussocky species, and creeping red fescue *F. rubra* and timothy *Phleum pratense* can be used to ensure good ground cover between these main plants. Less dominant and competitive species such as small timothy *P. bertolonii*, crested dogs-tail *Cynosurus cristatus*, smooth-stalked meadow grass *Poa pratensis* and Chewing's red fescue *F. rubra commutata* are better where a more open structure is required for foraging invertebrates or birds. Seed is generally sown at a rate of 20kg/ha.

Herb species can be incorporated into the mixture, but for optimal benefits to pollinators, a more targeted flower-rich sward should be established. Some flowering plants such as dandelion *Taraxacum officinale* and white clover *Trifolium repens* often colonise grass margins naturally, but others that might be incorporated include red clover *T. pratense*, birds-foot trefoil *Lotus corniculatus*, or knapweed *Centauria nigra*.

If the land is coming straight out of an arable crop, then light cultivation may be all that is necessary by way of preparation, but if there are weed species that might dominate the sown mixture, such as docks and thistles, especially if these species are present in the seed bank, then repeated cultivation and spraying with a broad-spectrum herbicide is likely to be necessary.

Sown species tend to be small-seeded, so a fine seedbed is essential for successful establishment. Seeds can be broadcast on the surface and then rolled, or if drilled, should not be to a depth greater than 1.5 cm. Warm moist conditions are needed, either in September, or in April, or later in the spring if conditions permit.

In the first year, it is important to cut and remove the developing sward three or four times to control rapidly growing annuals and give a competitive advantage to the sown perennials. Cutting will also encourage the grasses to tiller and form dense structures close to ground level.

Allerton Project research has investigated the ecological role of grass field margin strips and how they can best be managed to deliver multiple benefits such as biological control of crop pests. For example, spiders are one group of invertebrates which contributes to the control of flying crop pests such as aphids. Haughton (2000, 2001) investigated the effect of herbaceous field boundary vegetation structure on spiders by applying different rates of glyphosate herbicide to experimental field margin plots. While the herbicide itself was not directly harmful to spiders, glyphosate reduced the height and complexity of vegetation which had a considerable impact on spider numbers and on the composition of the spider community. Web-building spiders, mainly *Gonatium rubens* and *Lepthyphantes tenuis* which capture flying crop pests formed about 90% of the spider community in tall structurally diverse margins but were reduced to very low numbers when the vertical structure was lost, while wandering species were unaffected. Maintaining tall structurally diverse herbaceous vegetation maintains numbers of these important predators of crop pests.

Many Carabid and Staphylinid beetles use hedgerow vegetation as an overwintering habitat, and in the spring and summer, move out into the crop to feed on aphids and slug eggs. Amongst a total of 56 polyphagous beetle species in field boundaries at Loddington, Collins et al. (2003) recorded high densities of *Tachyporus hypnorum* and *Demetrias atricapillus*, two species known to be important predators of aphids, of up to $401m^{-2}$ and $274m^{-2}$ respectively.

To take advantage of the predatory activity of such species, beetle banks, low grassy banks developed by the Game & Wildlife Conservation Trust in the 1980s (Thomas et al., 1992), were created within the cropped area of fields at Loddington in the early 1990s and planted with a series of individual grass species plots for research purposes (Collins et al., 2003; Figure 5.2). These were found to support high numbers of beneficial predatory

---

Box 5.2 Beetle banks

The aim of creating a beetle bank is to reproduce the grass field margin habitat further into the field, but with the additional benefit of a bank to provide drainage. An earth bank is created that is approximately 0.4 metres high and 2 metres wide by two-directional ploughing. The length of this low bank will depend on the size of the field but a working gap of one sprayer width at each end allows the field to continue to be managed as a single unit.

The beetle bank is hand sown with a mixture of perennial grasses containing tussock and mat-forming species such as cocksfoot, timothy and creeping red fescue sown at $3g/m^2$. Tall growing wild flowers can be added to encourage other predatory groups such as hoverflies and parasitic wasps but these plants may reduce the dense grass area favoured by predatory beetles. The seed is sown either in the early autumn when the ridges are formed or in the following spring. It may be necessary to apply a broad-spectrum, non-residual herbicide to remove weeds.

Figure 5.2 Beetle bank across a sloping arable field at Loddington

The number of beetle banks needed per field will depend on its size and on the quality of existing boundaries. A square 16-hectare (40 acres) field will not need a beetle bank because insects should be distributed across the field in reasonable numbers. However, a 20-hectare field (50 acres), which has established boundaries with raised banks and an abundance of tussocky grasses will need one bank in its centre to achieve uniform cover in the early spring when the predators move into the crop.

Once the beetle banks are established the grasses sown tend to exclude most weeds. A sterile strip between the ridge and the crop about 1 m wide, rotovated or treated with a residual herbicide can minimise the risk of damage to the grasses on the bank after applying a grass weed herbicide to the crop. After the first year, the banks are left uncut to allow tussocks to form. Thistles *Cirsium* spp. and docks *Rumex* spp. may become established in the bank. These can be spot treated with a knapsack sprayer containing a suitable selective herbicide before they become abundant. Beetle banks require little additional management.

beetles, most notably a mean density of 327 Carabids per square metre in plots of Cocksfoot *Dactylis glomerata*. The taller grasses also supported Lycosid and Linyphiid spiders, each at densities of between 70 and 90 per square metre, while overwintering Staphylinid beetle densities often exceeded 1,000 per square metre. The effect of such high predator densities on numbers of aphids in adjacent crops was investigated, revealing that beetle banks were associated with reductions in aphid numbers at least 80 metres into the cropped area from the banks.

Beetle numbers in beetle banks fluctuated more than in field margins, and this may be due to the higher soil organic matter and botanical and structural diversity in field margins than in beetle banks. As *T. hypnorum* and *D. atricapillus* do not enter an obligate diapause (inactive state) in winter, they may benefit from the greater shelter and prey availability in field margin hedges than in beetle banks. As they overwinter as adults and breed in the spring they may more readily move out into crops in the spring, compared with those species which lay eggs in the autumn and overwinter as larvae. Field margin habitats also provide important stable soil conditions for overwintering and pupation. This role is discussed further in relation to hedges below.

Jones (2014) assessed earthworm numbers at intervals along eight 20-metre transects through grass margins and into adjacent crops. Earthworm densities were more than twice as high in grass margins as in the cropped land (Figure 5.3). In this case, the cropped area was under a reduced tillage system so we can expect a greater difference between grass margins and the cropped area where earthworm numbers are reduced by ploughing (see Chapter 3 for more on this issue). High earthworm numbers in grass margins have the potential to increase rates of water infiltration, reducing surface runoff to adjacent watercourses, and represent a refuge from which worms can colonise cropped land where worm numbers become impoverished as a result of cultivation.

Birds also benefit from the establishment of herbaceous field margins. Many farmland bird species are associated with field margin habitats where habitat structure influences the bird community present. The probability of whitethroat *Sylvia communis* breeding territory establishment in field boundaries is strongly associated with the width of herbaceous vegetation

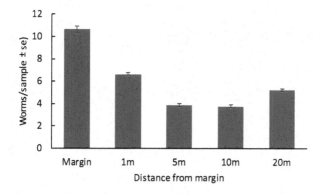

Figure 5.3 Earthworm abundance along a transect from a grass margin into an arable field (drawn from data in Jones, 2014)

in field margins, while tall hedges tend to reduce the use of field margins by this species (Stoate & Szczur, 2001; Figure 5.4). Yellowhammer *Emberiza citrinella* breeding territory establishment is influenced positively, both by the width of the herbaceous vegetation and by the presence of a ditch in field margins. Nest monitoring in the 1990s revealed that 95% of whitethroat nests and about half of yellowhammer nests were in the herbaceous field margin vegetation, rather than in the hedge itself and, for yellowhammers, nest predation rates were higher in hedges than in herbaceous vegetation. Whitethroats, on the other hand, nesting in field margin herbaceous vegetation, had the highest nest survival rates of all the bird species studied at Loddington. Our research therefore suggests that establishing herbaceous vegetation in field margins increases the nesting success and breeding abundance of these two bird species.

Bence et al. (2003) found significantly higher densities of harvest mouse *Micromys minutus* nests in beetle banks than in field margins, and in beetle banks, nests were built in the robust tussocky grasses such as cocksfoot which are favoured as overwintering habitat by predatory beetles. The vegetation was denser at harvest mouse nest sites than elsewhere in beetle banks. Murray (2004) found that skylarks *Alauda arvensis* and yellowhammers both selected nest sites with beetle banks within foraging range more than would be expected from the availability of this habitat in the landscape. Skylarks used beetle banks as a foraging habitat more than would be expected from their availability and were observed entering quite tall vegetation whereas yellowhammers tended to forage in more open vegetation in adjacent crops where they may have benefited from invertebrates dispersing from the beetle banks.

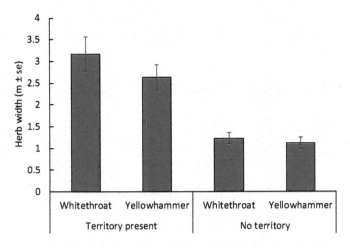

Figure 5.4 Width of herbaceous vegetation in field margins with and without whitethroat and yellowhammer breeding territories

Together, these various research projects highlight the benefits arising from creation of grass margins and beetle banks as a habitat for birds and mammals, and for earthworms, spiders and polyphagous predatory beetles which deliver ecosystem services linked to water and crop production. Grass margins against water courses also help to provide a buffer to reduce the impact of surface runoff and provide access for hedge cutting in later winter, after berries have been consumed by birds. As they mature, grass margins also become colonised by flowering plants which provide a resource for pollinating insects, although seed mixtures specifically designed for this purpose provide greater benefit.

Foraging habitats for pollinators

The use of flowering plants by wild bees and other pollinating insects differs between insect species, and seasonally between plant species, and with cutting regime. In an experimental plot study (Cresswell, 2018), the species most used by hoverflies were hogweed *Heracleum sphondylium*, ox-eye daisy *Leucanthemum vulgare*, red campion *Silene dioica*, black knapweed *Centaurea nigra*, yarrow *Achillea millefolium* and wild carrot *Daucus carota*. All have large floral display sizes and are attractive to a range of fly species. Of these plants, yarrow and wild carrot were not used by bumblebees, but their inclusion in seed mixtures extends the range of pollinating insects to benefit to include hoverflies and others. As well as being pollinators as adults, the larvae of hoverflies are also obligate aphid predators and can contribute to aphid control in crops. Of the other pollinator groups, beetles made significantly greater use of ox-eye daisy than other plant species.

Box 5.3 Flower-rich pollinator habitat

The objective of this habitat is to provide a mixture of native mainly perennial flowering plants which meet the requirements of as wide a range of bee, butterfly and other invertebrate species as possible throughout their foraging period, extending from the early spring when some adult insects emerge from hibernation to early autumn (Figure 5.5). The mixture therefore needs to contain several plant species. Early flowering species such as black medic *Medicago lupulina*, cowslip *Primula veris* and dandelion can provide important early-season foraging for bees where conditions are suitable for these species. Plant species known to grow locally can be used as a guide, and seed companies can often provide advice on the suitability of other species for site-specific characteristics such as soil type, fertility and moisture, aspect, shade, etc. Several plant species provide foraging

(Continued)

habitat for bees and other insects through the main summer period, and shrub and tree species provide earlier (e.g. willows *Salix* spp.) and later (e.g. ivy *Hedera helix*) sources of food for bees and other pollinating insects before and after hibernation.

Wherever possible, seed mixtures should consist of seed from UK sources. Seed of wild species is also available from other countries, and seed of agricultural cultivars is also available for some species, but they are not necessarily suitable for local pollinators or growing conditions. They tend to be cheaper but may not have the longevity required for this habitat type. Where there are nearby swards containing appropriate species, 'green hay' can be harvested and applied at a rate of 1ha of hay to 2ha of ground. The hay needs to be spread within a maximum of a day of harvest to avoid spoiling the seed contained within it as a result of increased temperature.

For the best establishment of flowering plants, only the less vigorous fine-leaved grasses are incorporated in the seed mixture, including such species as bents *Agrostis* spp., sheep's fescue- *Festuca ovina*, small timothy, crested dogs-tail, smooth-stalked meadow grass and Chewing's red fescue. This avoids the habitat degenerating into a grass-only sward. The seed rate for such mixtures is normally 20kg/ha, comprising a minimum of 10% wild flower species by weight. A higher proportion of wild flower species would be advisable if the seedbed is suboptimal.

If the habitat is being created on recently cropped arable land, a simple grass sward can be established and subsequently harvested, removed and destroyed to reduce the soil nutrient status before sowing the wild flower mixture. This helps give a competitive advantage to the sown species rather than vigorous grasses and perennial weeds. Where the ground already has these species as plants or as a seed bank, then repeated cultivation and application of a broad-spectrum herbicide reduces the risk of their dominating the sown sward. Avoiding shady areas or areas adjacent to vigorous perennial weeds that can spread into the newly created habitat, and maximising the area sown will also reduce this risk. Wider areas are also less prone to spray drift from arable land, whether of herbicides which would affect the sown plants, or insecticides which would affect insects foraging on them.

The seeds of sown species are small and require a fine seedbed. Seed is broadcast or shallow drilled and then rolled into a warm moist seedbed. April and September normally provide the best conditions for establishing these seed mixtures. The sward is then cut three or four times in the first year to reduce vigorous annual weeds and encourage perennial plants. While some species such as knapweed, birds-foot trefoil, self-heal *Prunella vulgaris*, yarrow *Achillea millefolium* and ox-eye daisy *Leucanthemum vulgare* establish quickly, others are slow to do so and may not appear until three years or more after establishment.

Across a range of plant species mixtures, plant cover increases gradually over a three-year period (Cresswell, 2018). The sward is best cut annually at the end of the summer, ideally with the cut material being removed to reduce the soil fertility, although practical constraints often make this difficult. Cutting and removal encourages the development of perennial species over annual plants which leave the bare ground when they die back each year. Alternatively, the sward can be grazed in the autumn. If grasses dominate the plant community, a low rate graminicide application in spring can reduce grass vigour, favouring the other plants, but this requires professional advice and is not permissible in all funded schemes. The inclusion of yellow rattle *Rhinanthus minor* in the seed mix reduces the competitive advantage of the grasses and creates a more open structure to the sward, as this species is parasitic on grasses and some other plants. Cutting part of the area in June can extend the flowering period but is not permissible under many agri-environment schemes. Cutting some areas only every two years can favour some species such as bush or tufted vetch *Vicia sepium and V. cracca.*

Figure 5.5 Flower-rich margin along the edge of a young woodland plantation

In terms of invertebrate crop pest predators, spiders were associated with a higher proportion of grasses in the same plant species mixture, especially tall upright species such as cocksfoot, while carabid beetles responded to a higher proportion of forbs (Cresswell, 2018). Eighty-seven percent of visits to flowers by parasitic and other wasps were to yarrow, so the presence of this species increases the diversity of crop pest predators and parasitoids present.

In 2014 and 2015, a broad group of pollinating insects was systematically surveyed in vegetation adjacent to small field margin ponds, and in a comparable length of the same field margin without a pond (John Szczur, unpublished data, 2015). Pond banks were sown with a wild flower seed mixture when the ponds were created in 2013. There were 6,582 records of pollinating insects at ponds and 3,202 at the control sites, but insect numbers were not consistently higher at ponds than at control sites. In 2014 and 2015 combined, the naturally occurring plant species willowherb *Epilobium* spp., thistles *Cirsium* spp. and umbellifers such as hogweed and cow parsley together accounted for 93% of the pollinator records.

In the control areas (without ponds and sown seed mixtures), a wide range of bumblebee species visited spear thistle *Cirsium vulgare* and great willowherb *Epilobium hirsutum*. Around the ponds, the sown species, wild carrot, bird's-foot trefoil, black knapweed and ox-eye daisy were the most visited species, and bird's-foot trefoil and tufted vetch were particularly frequently used by *Bombus pascuorum* and *B. lapidarius*. As at the control sites, bumblebees also visited spear thistle.

Wild flower seed mixtures also provided a more species diverse foraging habitat for butterflies, with species such as birds-foot trefoil and tufted vetch being the main plant species used, whereas in control sections, a similar number of butterflies was recorded but foraging was largely restricted to creeping thistle. These observations highlight the importance of naturally occurring plant species such as great willowherb, creeping thistle, and spear thistle to a range of pollinating insects, especially as sown species established only around the larger ponds in this study because of high bank soil fertility and competition with the more vigorous existing species at smaller sites. However, where they established, sown species increased the diversity of plants available as a foraging resource and extended the period in which flowering plants were available to foraging insects.

Some wild bird seed mixtures that are planted to provide winter food for birds (see below) also support pollinating insects in summer, with kale *Brassica oleracea* and yellow melilot *Melilotus officinalis* being recorded as the main plant species supporting bees in one study at Loddington (Tomlin, 2018). Kale provided the foraging resource early in the summer, with melilot flowering later in the season. In this case, wild bird seed mixtures containing these two species supported around twice the bee density through the summer as nearby hedges, wildflower mixtures or other types of wild bird seed mixture.

Hedge management

Hedges have been a traditional component of lowland farmland landscapes since the Enclosure Acts of the 17th century, but many were removed in the second half of the 20th century. More recently, there have been numerous national government and local authority incentives to re-introduce hedges into the landscape, although their role is now less for livestock management and more to meet wildlife conservation and aesthetic landscape objectives (Figure 5.6). Hedge creation and management therefore differ in some respects from that of the early plantings. For example, the number of shrub species planted is higher and hedges are encouraged to grow taller. This benefits most species, although others such as whitethroat favour lower hedges (Stoate & Szczur, 2001) and while planting up gaps creates a more continuous habitat, gaps can be important for the movement of some wildlife between fields (e.g. Lepidoptera, Dover, 2019). Hedge planting and increasing the volume of hedges also increase the potential for carbon storage, both above and below ground.

Hedges represent a woody habitat through farmland, but the structure is very different from that of woodland, herbaceous vegetation is less shaded, and the species composition of both woody and herbaceous plant communities may also be different, while because of their narrow width, hedges may be influenced more by adjacent farmland than woodland in the same landscape. Despite this, both hedges and woodland edges represent stable habitats in which insects can complete all or part of their life cycles.

While they support many woodland species, hedges differ from woodland edges in the invertebrate species that they support, and the frequency of cutting, and the resulting height and structure of the hedge influence the ecology of this habitat. As described in Chapter 4, less frequently cut hedges are associated with a greater abundance of flowers, pollinating insects, and production of fruit which forms a source of food for some bird species in winter.

Figure 5.6 Flowering hawthorn hedge with associated tall herbaceous vegetation including flowering red campion *Silene dioica*

Hedges are widely valued as a habitat for birds in the breeding season when the structure of the hedge can influence nest survival. In a landscape scale study involving 399 nests in 2003 and 2004, nests in infrequently managed hedges that were consequently open structured were associated with higher failure rates (82.4%) than nests in hedges that were managed to maintain a dense structure (32.7%; Dunn et al., 2016). This is discussed further in Chapter 7.

Box 5.4 Hedge management

The overall aim of hedge management should be to achieve a range of hedge heights and structures across the farm as different wildlife species have different requirements, and to have hedges at a range of management stages at any one time. This requires a formal hedge management plan so that it is clear which hedges are to be managed in which way in any one year. Taller hedges generally support higher wildlife species diversity than short ones but exclude some species.

Traditional laying of hedges on rotation every eight to twenty years is beneficial in maintaining hedge structure. Hedges can be cut using flail cutters, finger bar cutters or circular saws. Flail cutters are by far the most commonly used, and although often criticised for producing an untidy appearance, have not been demonstrated to have any detrimental effect on the long-term viability of hedges.

Hedges are best cut every other year, or every three years. No more than a third of hedges should be cut in any one year. This is to ensure that there are lengths of hedge that are sufficiently mature to provide berries as food for birds and to support overwintering invertebrates. Hedges are best cut in the second half of the winter when berries are no longer present. This is often not possible because the adjacent land is cultivated or ground conditions are too wet to allow the use of heavy machinery. Grass or wild flower strips along field margins therefore allow late winter hedge cutting to take place. Where it is necessary to cut hedges annually, associated habitats such as grass margins can be optimised to benefit those species that are associated with low hedges.

Many mature shrubs provide berries, nest sites and year-round foraging habitat for a wide range of species where these taller sections will not adversely affect productive land. Other mature trees provide similar benefits. Mature trees and tall lengths of hedges also provide shade and shelter for livestock, but there can be associated compaction and eutrophication of field margins through deposition of dung and urine. Similarly, livestock benefit from browsing hedge and tree species, but this should not be to the detriment of the hedge. Livestock can be fenced away from hedges if they are likely to damage them through excessive browsing, destruction of the hedge base vegetation, or compaction of the ground.

Broken lengths of hedges containing many gaps are often 'gapped up' with new planting of hedgerow shrubs. This increases the shrub area and hedge length but may reduce the suitability for some species. Where there is not an abundance of herbaceous vegetation in field margins, hedge gaps often provide the only site for this habitat, and hedge gaps can be important for the movement of flying invertebrates such as butterflies and moths across landscapes in which hedges represent barriers to movement.

Winter food for birds

Reduced winter survival rates of birds are thought to have driven national population declines of some species, largely as a result of depleted sources of seed food during winter (Siriwardena et al., 2008). By experimentally investigating the use of a range of annual and biennial seed-bearing crops by a range of farmland bird species, we were able to design seed mixtures that would best provide winter seed food (MAFF, 2001; Stoate et al., 2003, 2004).

Greenfinch *Carduelis chloris* was generally the most abundant observed species and differed from most other species in its feeding preferences. Among the annual plants, feeding of greenfinches was largely confined to four crops, borage *Borago officinalis*, sunflower *Helianthus sp.* mustard *Sinapis alba* and linseed *Linum usitatissimum*, but both the numbers of birds and relative use of these crops changed as the winter progressed. In September, greenfinches fed on borage and sunflower, but as the borage seed was soon depleted, feeding became more concentrated on sunflower during October and November. By December, sunflower seeds were also becoming depleted, and feeding switched to mustard during December and January. Small numbers of birds were observed feeding on linseed throughout the winter.

Chaffinches *Fringilla coelebs* also used a range of crops but feeding was mainly confined to the closely related fat hen *Chenopodium alba* and quinoa *Chenopodium quinoa*. Goldfinches *Carduelis carduelis* also used several crops and significant differences between crop types were only detected in one year when most birds fed on linseed. For the buntings, reed buntings *Emberiza schoeniclus* and yellowhammers, there were significant differences between crops. Reed buntings used a variety of small-seeded crops, especially millet and fat hen. Yellowhammers fed almost exclusively on the cereals – wheat *Triticum sp.*, triticale *Triticum x Secale* and millet *Panicum miliaceum*.

Among the biennial crops, greenfinch and chaffinch were most abundant in kale *Brassica napus*. In contrast, goldfinches rarely used kale, preferring teasel *Dipsacus fullonum*, and to a lesser extent, evening primrose *Oenothera biennis*. The largely insectivorous blackbird *Turdus merula* and song thrush *Turdus philomelos* used kale and to a lesser extent chicory *Cichorium intybus*, probably as cover and a source of invertebrate food associated with moist soil conditions, more than for any seed food benefit provided by the crops (Figure 5.7).

We monitored seed availability throughout the winter, revealing substantial depletion to mid-winter when seed abundance in most plots was exhausted. As a result, bird numbers using the plots declined sharply in mid-winter (Figure 5.8). The exception was skylark which was only present in small numbers, but increased its use of crop plots from November to January, possibly because the structure became more open and suitable for this species to forage in.

As a result of this research and subsequent observations, wild bird seed mixtures were adopted as a habitat option in UK government agri-environment schemes, encouraging farmers to establish seed-bearing crops on their own land to provide seed food for farmland birds in winter. Such crops are most likely to be planted along field edges because of the lower impact on commercial crop production (Figure 5.6). By comparing the use by birds of seed-bearing crop plots planted along field margins with plots of the same crops planted in field centres, we were able to assess the impact of location on their use by birds. There was a significant effect of plot location on the abundance of blackbird, chaffinch, greenfinch and linnet. In each

Figure 5.7 Wild bird seed crop along a field margin at Loddington

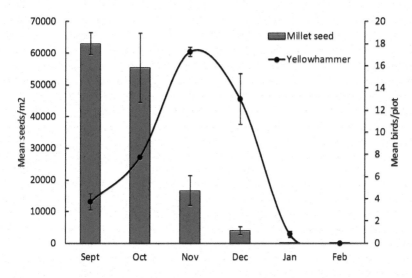

Figure 5.8 Changes in abundance of white millet seed in experimental plots and numbers of yellowhammers recorded using the same plots

case, abundance was higher in field margin plots than those away from the field edge.

Because seed food in wild bird seed crops becomes depleted by the middle of the winter, supplementary provision of seed food can fill this 'hungry gap', maintaining bird numbers through late winter and into the spring, and increasing their survival rates. Analysis of yellowhammer winter foraging observations revealed a switch from cereal-based wild bird seed crops to pheasant feed hoppers as a source of food from early to late winter (Stoate & Murray, 2002).

The role of this supplementary source of grain could be verified at Loddington through long-term monitoring of bird numbers. Grain was fed to pheasants from 1993 to 2006, both by scattering on the ground and from feed hoppers distributed around the farm. From 2007 to 2010, winter feeding was stopped to assess the impacts on those bird species known to use feed hoppers and then restored again from 2011. In order to understand which species were using feed hoppers, we used cameras positioned at feed sites. Approximately a third of the feed hopper use was attributed to gamebirds, a third to small birds, and a third to other species. Songbird species were predominantly chaffinch, yellowhammer, tree sparrow *Passer montanus*, blackbird, dunnock *Prunella modularis* and robin *Erithacus rubecula*.

Winter abundance of these species across the farm did not differ in the first half of the winter between years with and without supplementary feeding, presumably because birds were finding alternative food sources such as that available from wild bird seed crops. However, in late winter, bird numbers were more than twice as high when supplementary food was provided. More importantly, abundance of the hopper-feeding bird species in the late winter (February and March) was correlated with breeding abundance in the following spring, with a 30% increase in breeding bird numbers when supplementary food was provided (Figure 5.9). This is in line with results reported by Hinsley et al. (2010) who found that breeding numbers of birds increased in years following provision of wild bird seed crops, although in their case, in the absence of supplementary food provision.

As a result of feeding at the start of the day, birds rapidly lay down fat which increases during the day and provides energy during long winter nights and periods of food shortage such as the late winter period. Supplementary food at this time of year therefore reduces the movement of birds from individual farms and can be expected to increase the survival of birds, as well as the condition of birds as they enter the breeding season. However, some consideration needs to be given to the potential for also benefiting species such as brown rat *Rattus norvegicus* (Md Sad et al., 2020) that predate the species targeted for conservation, and for changing bird species community structure (Shutt et al., 2021).

During the breeding season, wild bird seed crops also provide a role as a foraging habitat because they are open structured and support invertebrates that are a source of food for birds. Both skylarks and yellowhammers used wild bird seed mixtures but skylarks used kale-based mixtures more than expected from their availability, while yellowhammers used cereal-based mixtures (Murray, 2004; Murray et al., 2002). Skylark chick growth rates were positively associated with the area of these crops around the nest, and negatively with the proportion of other habitats, demonstrating a breeding season benefit of this habitat.

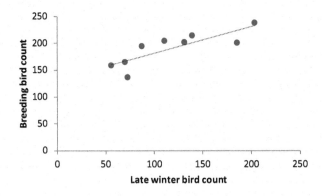

Figure 5.9 Late winter (February/March) abundance of hopper-using songbird species in relation to the abundance of the same species in the following breeding season

Box 5.5 Wild bird seed mixtures

Wild bird seed crops exploit the existing skills and equipment on any arable farm and these crops should be considered like any other crop in the rotation, except that inputs will be lower. Siting crops along hedges increases the use by birds as the adjacent cover provides protection from avian predators such as sparrowhawks.

A clean seedbed is important to avoid weed infestation, especially by perennial weeds such as docks and thistles on ground that has been used for wild bird seed crops for many years. A wide range of seed-bearing crops can be grown but even simple mixes of cereals and brassicas can meet the requirements of bird communities on individual farms. Red and white millets, canary seed, linseed, quinoa, sunflower and others can be added to increase diversity and spread the risk of any one component of the mixture failing to establish. As seed size varies, small-seeded species such as brassicas and millets are best broadcast on the prepared seedbed and the larger seeded species such as wheat or triticale are then drilled using a conventional seed drill and rolled. Seed companies provide guidance on details such as row width and drilling depth, according to the mix of species being used.

Establishing the crop is best done in the spring for most species, but autumn establishment, mainly of cereal-based mixtures, is also an option. These will provide seed food for birds in their second winter so need to be left in the ground for longer than similar spring-sown mixtures. Biennial mixtures based on kale can also be adopted, incorporating annuals to provide seed food in the first year, the kale providing the seed in the second.

For spring sowing, a previous crop needs to be destroyed sufficiently in advance to enable a good seedbed to be created. This presents a dilemma if that crop is still providing a source of food for birds, but this is rarely the case. Developing a rotation of wild bird seed crops across the farm creates the greatest flexibility, helps with disease management (e.g. clubroot in brassicas), and provides an opportunity for weed control using herbicides if necessary. Rotating broad-leaved and cereal-based mixes can help with weed control.

Adequate soil fertility is essential to achieving a good yield of seed for birds, just as it is for commercial crops. Fertiliser rates are lower for wild bird seed crops, but it is still important to monitor crop performance, test soils periodically and apply fertiliser as necessary.

Flea beetle can be a major problem for crop establishment in spring because of the severe damage caused to the very young plants. A seed dressing can help to reduce damage but the dressed seed must be drilled, and not broadcast on the surface. The use of insecticides should

(Continued)

be minimised to avoid impacts on non-target groups and, within Stewardship schemes, is permissible only if a derogation is obtained.

As an alternative to wild bird seed mixtures, especially in livestock systems where equipment and expertise might not be available for establishing crops, existing ryegrass can be allowed to set seed which provides a source of seed food, especially for buntings, through the winter. For this, grazing and cutting ryegrass are stopped at the end of May and the grass is left to flower and set seed. As with wild bird seed mixtures, areas alongside hedges are the most suitable. Grass containing a high proportion of clover is best avoided as the late summer clover growth can swamp the grass seed heads which then rot. Also, as with wild bird seed mixtures, seed will be produced only where there is sufficient fertility in the soil, so repeated use of the same area of land without the application of fertiliser will fail to produce an adequate source of seed food.

Box 5.6 Supplementary winter feeding

Seed food can be provided for birds throughout the winter but is most crucial in the second half of the winter when other sources of food have become depleted. Seed can be provided by regular scattering on bare ground or by the provision of feed hoppers (Figure 5.10) that represent a more constant supply with less effort. However, regular frequent provision of seed by scattering on open ground in the early morning increases the chance of seed being eaten by birds, reducing the risk of attracting rats which feed mainly at night.

If feed hoppers are used in addition to scattering, it is important to move them periodically through the winter to reduce the seed lost to rodents and avoid encouraging the population increase of mammal species which may have detrimental impacts on birds. Moving feed hoppers also reduces the risk of disease such as Trichomonosis which can be transmitted at regularly used feed sites.

Feed hoppers should be sited close to cover such as hedges or woodland to provide protection for feeding birds from avian predators such as sparrowhawks *Accipiter nisus*. Using a large number of feed hoppers distributed across the farm reduces the concentration of feeding birds, disease spread, and risk of predation.

A combination of the readily available seeds, wheat and oilseed rape meets the requirements of all bird species likely to be using feed hoppers in winter and avoids the need to buy seed from off-farm. However, current Stewardship scheme options require the purchase of a wider range of often imported seeds such as millet and canary seed.

Figure 5.10 Songbird feed hopper

Small-scale wetlands

As well as the historical loss of sources of winter food from farmland land-scapes, there has also been a loss of wetland habitats in the form of field ponds, bogs, marshland and seasonally wet ground, largely as a result of field drainage. Increasingly dry summers associated with climate change can be expected to accentuate this trend with increasing loss of small wet-lands and wet areas of land. Our work in southern Portugal has shown that the presence of water is an important influence on bird abundance in these already dry landscapes (Borralho et al., 2000). Impeding field drainage sys-tems to restore these habitats would have a negative impact on crop produc-tivity and increase negative impacts on water, but an alternative is to create small field edge and field corner ponds outside the productive area.

In 2004, we created 37 small field margin ponds to assess the potential ben-efits to biodiversity and for capturing sediment from arable land over the sub-sequent four years (Defra, 2007). All the ponds were fed by water running off arable land or pasture via field drains and ditches, field drain outfalls being avoided to ensure that there were no impacts on adjacent productive land.

Most of the ponds were created simply by building an earth dam across a ditch (Figure 5.12), while six consisted of pairs of small ponds in field corners. We monitored the use of the ponds by birds by making a series of observations from a portable hide, and we monitored emergent aquatic insects using a set of five floating emergence traps at each site. In each case, we gathered data both from the ponds and from paired sections of field margin without ponds.

The creation of these small ponds, which held water into the summer, resulted in the presence of 10–32 aquatic invertebrate species, including six nationally scarce beetles.[1] They were also used by Common Frog *Rana temporaria*, Smooth Newt *Triturus vulgaris* and Great Crested Newt *T. cristatus* and were associated with up to eight aquatic plant species (Aquilina et al., 2007; Freshwater Habitats Trust results in Defra, 2010).

As well as providing benefits to farmland wildlife, these features can also trap sediment eroded from adjacent land. While the resulting exposed mud can provide a habitat for insects, accumulated sediment will need to be removed periodically and spread on the field.

Emergent aquatic insect biomass per unit area was greatest where there were more extensive areas of bare wet mud and lower levels of shading from hedges. As the area of water and mud was also greater in ditch sections with earth dams than in control sections, the absolute biomass of insects associated with dammed sections was approximately four times that of control sections. Initially, there was a tendency for ponds in pasture catchments to have higher insect biomass than those in arable catchments, but there was less of a difference after four years. Emergent insect biomass in unmanaged ditches declined across the four years of the study, but where we had removed accumulated silt from some ditches in the third year, emergent insect biomass increased again in those ditches in the following year (Freshwater Habitats results in Defra, 2010).

We found that a wide range of bird species used the ponds for foraging, drinking, bathing and provisioning young. Ponds were associated with significantly more bird visits than control sections of the same field boundary, especially in summer and autumn when there were approximately twice as many visits to ponds as to control sections of field margin (Figure 5.11). Bird visit rates were initially positively correlated with aquatic insect biomass production, suggesting that birds were actively seeking out areas where foraging would be most productive, but this was not maintained in the latter part of the project when the difference in insect biomass between ponds and control sections of field margin was lower.

Although these ponds were receiving runoff from agricultural land, resulting in elevated sediment and nutrient concentrations and eutrophication of the habitat, these results illustrate the contribution that small ponds can make to biodiversity on farmland at the landscape scale. It is important to note though that the biodiversity benefits do not compare with those resulting from the creation of clean water ponds which do not receive agricultural runoff.

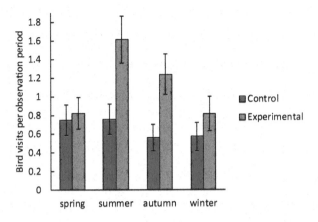

Figure 5.11 Relative use of bunded and control sections of ditches by birds throughout the year

Box 5.7 Bunded ditches

The simplest means of creating small-scale wetland habitat on grass
or arable farms located on clay soils is the creation of earth dams in
ditches (Figure 5.12). These are not appropriate for very low-lying areas

Figure 5.12 Bunded ditch at Loddington

(*Continued*)

and dams should be kept well clear of field drain outlets to avoid waterlogging within the adjacent field, but otherwise, their creation has
no impact on the productive area. The ditch can be widened slightly
upstream of the dam, providing clay material for constructing the dam
itself. The dam is keyed into the ditch base and bank at each side to
reduce the risk of erosion during high flows. It is well compacted and
contains a pipe of an appropriate diameter to accommodate normal
flow close to the top of the dam. The top of the dam should be slightly
lower than the level of the adjacent land in case excessive flow exceeds
the capacity of the pipe.

   Larger ponds can be created in unproductive field corners. In this
case, a dam can be created in an adjacent ditch to divert water into
the pond at peak flow, or the pond can simply receive surface runoff
from the field.

Landscape scale perspectives

In 2007, four of the simple field margin ditch ponds referred to above were
surveyed for zooplankton and larger invertebrates along with three small
ponds within arable fields and three small ponds in low-input pasture
fields (de Bie et al., 2010). This cluster of ponds with differing land-use
micro-catchments formed part of a wider European study involving eight
other pond clusters in Belgium, Denmark, Germany, Poland, Slovenia and
Hungary. With the exception of the large temporary ponds of Hungary, the
relationship between pond ecology and the adjoining land use was broadly
similar across Europe. Ponds that were close to arable land tended to be
characterised by high values of water turbidity, total phosphorus, chlorophyll *a* concentrations, and by sparser aquatic vegetation. The most important land-use effects on ponds operate at relatively small rather than
at large spatial scales, and inputs of nutrients mainly originate from local
surface water runoff rather than from atmospheric deposition or major
groundwater flows. However, across Europe, landscapes that are dominated by arable land could also be characterised by low pond densities,
with pond isolation reducing opportunities for colonisation. Most of the
zooplankton species found in our ponds are widely distributed and present across most of Europe. The distribution of several macroinvertebrates
is more restricted and this might explain why 'country' as a variable explained more of the variation in macroinvertebrates than in zooplankton.

   With their small volumes, ponds are highly vulnerable to degradation
caused by surface water pollution derived from their immediate surround-
ings. Unlike lakes and rivers, there is little possibility of dilution or buffering
of pollutant inputs. Consequently, poor-quality ponds are often degraded to
an extreme degree that is rarely seen in larger waters. However, the converse

is that the creation of new ponds within grassland micro-catchments has considerable potential for increasing aquatic biodiversity at the landscape scale. We explored this potential from 2013 in our landscape scale Water Friendly Farming project and discuss the results in the context of catchment management objectives in Chapter 6.

Adjacent land use and proximity of other habitats will also influence the establishment and ecological function of terrestrial habitats. High nutrient status soil and shading from woodland can have negative impacts on the establishment and use of pollinator habitats for example. Dispersal and foraging ranges are another important consideration. In winter, birds are highly mobile and even quite dispersed wild bird seed mixtures can be exploited (Calbrade et al., 2006), but in the breeding season, their movements are constrained by nest location. Our research identifies breeding season foraging ranges of up to 300 metres for yellowhammer, up to 200 metres for tree sparrow and skylark, and up to 120 metres for song thrush. If wild bird seed mixtures or other habitats are to be used by a large proportion of the local bird population in the breeding season, their distribution needs to reflect these distances.

Davis et al. (2010) used a Geographical Information System to investigate the accessibility of wild bird seed mixtures at Loddington and adjacent farms to birds in winter and during the breeding season. Through the adoption of agri-environment schemes on local farms, the distribution of wild bird seed crops in the landscape was sufficient to provide access to birds in winter across most of the area. However, based on foraging distances of 200m and 300m from randomly generated 'nest sites' across the same landscape, the crops were accessible to only 15% and 21% of nests respectively.

For species such as dunnock and whitethroat, the foraging range is only a few metres from the nest, so this is unlikely to influence decisions about distribution of wild bird seed crops across the farm. Maximum foraging ranges of many bumblebees and solitary bees are 150–600 metres (Darvill et al., 2004; Gathmann & Tscharntke, 2002), and so a distance of this magnitude can be used as a guide for the creation of nesting and foraging habitats. Carabid beetle dispersal distances between overwintering and foraging areas range from up to 60 metres to several hundred metres, depending on species (Thomas et al., 2002). Spatial distribution of habitats at farm and landscape scales is therefore an important consideration, both in terms of the influence of adjacent habitats and the applicability to the life cycle and ecology of target species.

Note

1 *Agabus chalconatus, Berosus signatocolis, Haliplus heydenni, Hydraena nigrita, Hydrglyphus pusillus* and *Rhantus suturali*s.

References

Aquilina, R., Williams, P., Nicolet, P., Stoate, C. & Bradbury, R. (2007) Effect of wetting up ditches on emergent insect numbers. *Aspects of Applied Biology* 81, 261–262.

Boatman, N. D., Blake, K. A., Aebischer, N. J. & Sotherton, N. W. (1994) Factors Affecting the Herbaceous Flora of Hedgerows on Farms and Its Value as Wildlife Habitat. In: Watts, T. A. & Buckley, G. P. (Eds.) *Hedgerow Management and Nature Conservation.* Wye College Press, Wye. 33–46.

Borralho, R., Stoate, C. & Araújo, M. (2000) Factors affecting the distribution of Red-legged Partridges *Alectoris rufa* in an agricultural landscape of southern Portugal. *Bird Study* 47, 304–310.

Calbrade, N. A., Siriwardena, G. M., Sutherland, W. J. & Vickery, J. A. (2006) The effect of the spatial distribution of winter seed food resources on their use by farmland birds. *Journal of Applied Ecology* 43, 628–639.

Chaney, K., Wilcox, A., Perry, H. H. & Boatman, N. D. (1999) The economic so of establishing field margins and buffer zones of different widths in cereal fields. *Aspects of Applied Biology* 54, 79–84.

Collins, K. L., Boatman, N. D., Wilcox, A. & Holland, J. M. (2003) A 5-year comparison of overwintering polyphagous predator densities within a beetle bank and two conventional hedgebanks. *Annals of Applied Biology* 143, 63–71.

Cresswell, C. J. (2018) *Multifunctional field margin vegetative strips for the support of ecosystem services – pollination, bio-control and water quality protection.* Unpublished PhD thesis. Harper Adams University.

Darvill, B., Knight, M. E. & Goulson, D. (2004) Use of genetic markers to quantify bumblebee foraging range and nest density. *Oikos* 107, 471–478.

Davis, F., Ewald, J. A. & Stoate, C. (2010) The influence of wild bird seed mixture spatial distribution at the landscape scale on conservation potential to farmland birds in summer and winter. *Aspects of Applied Biology* 100, 433–436.

De Bie, T., Stoks, R., Declerck, S., De Meester, L., Van De Meutter, F., Martens, K. & Brendonck, L. (2010) Agricultural Land Use Shapes Biodiversity Patterns in Ponds. In: Settele, J., Penev, L., Georgiev, T., Grabaum, R., Grobelnik, V., Hammen, V., Klotz, S., Kotarac, M. & Kühn, I. (Eds.) *Atlas of Biodiversity Risk.* 226–227. Pensoft, Sofia.

Defra (2010) *Wetting Up Farmland for Biodiversity. Research Project Final Report BD1323.* Department for Environment, Food and Rural Affairs, London.

Dover, J. W. (2019) *The Ecology of Hedgerows and Field Margins.* Routledge, Oxon.

Dunn, J. C., Gruar, D. Stoate, C., Szczur, J. & Peach, W. (2016) Can hedgerow management mitigate the impacts of predation on songbird nest survival? *The Journal of Environmental Management* 184, 535–544.

Gathmann, A. & Tscharntke, T. (2002) Foraging ranges of solitary bees. *Journal of Animal Ecology* 71 (5) 757–764. DOI:10.1046/j.1365–2656.2002.00641.x

Haughton, A. J. (2000) *Impact of glyphosate drift on non-target field margin invertebrates.* Unpublished PhD thesis. Open University.

Haughton, A. J., Bell, J. R., Boatman, N. D. & Wilcox, A. (2001) The effects of the herbicide glyphosate on non-target spiders: Part ll. Indirect effects on *Lepthyphantes tenuis* in field margins. *Pest Management Science* 57 (11), 1037–1042.

Hinsley, S. S., Redhead, J. W., Bellamy, P. E., Broughton, R. K., Hill, R. A. S., Heard, M. S. & Pywell, R. F. (2010) Testing agri-environment delivery for farmland birds at the farm scale: the Hillesden experiment. *Ibis* 152, 500–514. DOI:10.1111/j.1474–919X.2010.01029.x

Jones, S. (2014) *Environmental Stewardship grass margins as a source of earthworm colonisation for degraded arable soils.* Unpublished MSc thesis. University of Nottingham.

MAFF (2001) *Designing crop/plant mixtures to provide food for seed-eating farmland birds in winter.* Final Project Report BD1606. Ministry of Agriculture, Fisheries and Food, London.

Md Saad, S., Sanderson, R., Robertson, P. & Lambert, M. (2021) Effects of supplementary feed for game birds on activity of brown rats Rattus norvegicus on arable farms. *Mammal Research* 66, 163–171. DOI:10.1007/s13364-020-00539-2

Murray, K. A., Wilcox, A. & Stoate, C. (2002) A simultaneous assessment of skylark and yellowhammer habitat use on farmland. *Aspects of Applied Biology* 67, 121–127.

Murray, K. A. (2004) *Factors affecting foraging by breeding farmland birds.* Unpublished PhD thesis. Open University.

Shutt, J. D., Trivedi, U. H. & Nicholls, J. A. (2021) Faecal metabarcoding reveals pervasive long-distance impacts of garden bird feeding. *Proceedings B is the Royal Society's* 288, 20210480. DOI:10.1098/rspb.2021.0480

Siriwardena, G. M., Calbrade, N. A. & Vickery, J. A. (2008) Farmland birds and late winter food: does seed supply fail to meet demand? *Ibis* 150, 585–595. DOI:10.1111/j.1474–919x.2008.00828.x

Stoate, C. & Szczur, J. (2001) Whitethroat Sylvia communis and Yellowhammer *Emberiza citrinella* nesting success and breeding distribution in relation to field boundary vegetation. *Bird Study* 48, 229–235.

Stoate, C. & Murray, K. (2002) A new design for the arable landscape and its use by farmland passerines. In Chamberlain, D. & Wilson, A. (Eds) *Avian Landscape Ecology* 342–345. International Association for Landscape Ecology (UK)

Stoate, C., Szczur, J. & Aebischer, N. J. (2003) The winter use of Wild Bird Cover crops by passerines on farmland in North East England. *Bird Study* 50, 15–21.

Stoate, C., Henderson, I. G. & Parish, D. M. B. (2004) Development of an agri-environment scheme option: seed-bearing crops for farmland birds. *Ibis* 146 *suppl.* 2, 203–209.

Szczur, J. (2015) Unpublished data.

Thomas, M. B., Wratten, S. D. & Sotherton, N. W. (1992) Creation of 'island' habitats in farmland to manipulate populations of beneficial arthropods: predator densities and species composition. *Journal of Applied Ecology* 29, 524–531.

Thomas, C. F. G., Holland, J. M. & Brown, N. J. (2002) The Spatial Distribution of Carabid Beetles in Agricultural Landscapes. In: Holland, J. M. (Ed.) *The Agroecology of Carabid Beetles.*

Thompson, G. A. (2013) *An evaluation of the extent to which semi-natural habitats enhanced through agri-environment schemes support carabid beetles.* Unpublished MSc thesis. University of Nottingham.

Tomlin, F. (2018) *Do pollinators use bird focused agri-environment scheme plots?* Unpublished MSc thesis. Warwick University.

6 Reflections on water

Sediment and aquatic ecology

While farmland wildlife conservation needs to be considered in an integrated way, and at a landscape scale, that is even more the case for aquatic biodiversity and other environmental objectives associated with water. The way land is managed largely determines the quality of water reaching aquatic species. As Chapter 3 describes, land management has a particular influence on the movement of sediment and associated nutrients from arable land to water through surface runoff and sub-surface flow. In lowland clay catchments, field drains are thought to be responsible for up to 55% of the suspended solid load (Russell et al., 2001).

Sand and silt play an important role in aquatic ecosystems, providing a stream bed substrate on which many macrophytes and invertebrates depend. However, elevated sediment concentrations in water and deposition of finer material within channels have a negative effect on many species (Hauer et al., 2018). Brown trout *Salmo trutta* require a stream substrate with no more than 12% of sediment with a diameter of 1 mm (Crisp & Carling, 1989) as larger amounts of fine material reduce oxygen availability to eggs and impede emergence of newly hatched alevins.

Suspended sediment is defined as the material of <63 μm as particles of up to this diameter can be held in suspension in slowly moving water (Owens & Walling, 2002). Species particularly affected are those with gill respiration, including fish, but also invertebrates such as Plecoptera (stoneflies). Turbidity associated with suspended sediment also has a negative impact on visual feeders such as trout and salmon.

In 2004, we used a standardised quantitative sampling technique to survey wild brown trout at ten sites along the length of the Eye Brook. Age classes were separated by length into 'fry', 'one-year-old', and 'older' trout and we estimated the density of fish per 100 m^2 using the site area and numbers of fish caught. Brown trout were recorded at every site and densities of trout fry, one-year-old, and older trout ranged between 0.0 and 65.8, 0.6 and 32.2 and 0.0 and 6.3 per 100 m^2, respectively (Figure 6.1). The highest densities of trout fry were recorded in the headwaters because, as in other rivers of this type,

DOI: 10.4324/9781003160137-6

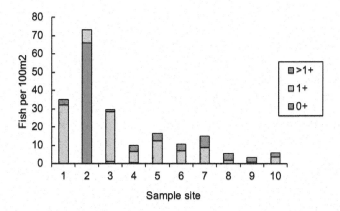

Figure 6.1 Brown trout density at Eye Brook sampling sites from the headwaters (1)
to the base of the catchment (10) for young fish (0), one-year-old fish (1+),
and older fish (>1+)

the most suitable spawning habitat is in the upper reaches and smaller tributaries. The largest numbers of one-year-old trout were also recorded in the
upper reaches. Again, as is found in other rivers, the highest densities of older
trout were recorded in the lower parts of the catchment. Similar results were
obtained in 2008 when Eye Brook tributaries were surveyed as well as three
of the mainstream sites surveyed in 2004 (Stoate & Sandford, 2008). These
survey results confirm that breeding success and recruitment are low, and
this is largely attributable to the sedimentation of potential spawning gravels.

Although it has the greatest ecological impact in stream riffles, sediment
is unevenly distributed across the stream bed, being concentrated in the
pools. In streams near Loddington, Nicholls (2015) recorded material of
less than 2 mm diameter comprising $40 \pm 3\%$ of the total sediment present in
pools and $28 \pm 2\%$ in riffles. The slower flow in pools enables finer material
to accumulate there, but it is also more readily remobilised by the current.
Fines remobilised when the water column in pools was disturbed amounted
to 367 ± 94 g m^{-2}, compared with just 38 ± 13 g m^{-2} from riffles. However,
even in the pools, the very fine clay fraction (<2 µm diameter) of the sediment was just 2%. This is considerably lower than the values of 20–60% recorded in catchment soils (Hamad, 2018), suggesting that at high flow, most
eroded clay material is held in suspension without settling so that it would,
in theory, be exported from the river basin to coastal waters.

In the case of the Eye Brook, Eyebrook Reservoir which is located in the
lower catchment, close to the stream's confluence with the River Welland
captures much of the material exported from the upper catchment. An investigation into reservoir bed sediment found that particle diameter was
predominantly in the 1–100 µm range (Ian Foster et al., unpublished, 2008).

Sediment size frequency distribution was predominantly bimodal in both the stream (Nicholls, 2015) and the reservoir (Foster et al., 2008), but in the reservoir, the peaks revealed much finer material (1–10 and 10–100 μm) compared to the stream (10–100 and 100–1,000 μm). The proportion of particles in the upper end of the size range was higher in the upper 15 cm of reservoir bed sediment cores, suggesting an increase in deposition of slightly coarser material since the early 1990s (based on radionuclide dating).

On average, sedimentation rates in the reservoir had increased by a factor of around three since the 1940s. The highest rates (approaching $0.5 \, \text{g cm}^{-2} \, \text{yr}^{-1}$) occurred in the late 1990s and early years of the 21st century. This trend is consistent with many lowland agricultural environments in England (Foster, 2006) and is associated with land-use change from pasture to arable in the 1940s. This is reflected in biological material in the upper bed sediment cores, with an increase in cereal pollen grains and a decline in pollen from pasture grasses and fungal spores from *Sporormiella*- and *Tripterospora*-type dung fungi associated with grazing livestock.

However, there is not a simple relationship between land use and sediment transport to water as topography, livestock access to streams and cultivation methods all have an influence. For example, Mitchell (2010) surveyed bed sediment in 18 tributaries of the Eye Brook and found higher levels of fine sediments in some pasture catchments than some arable ones. Limited application of geochemical analysis to the reservoir sediment data suggested an increase in sediment derived from subsoil in recent years, implying more severe soil erosion such as the creation of gullies or increased erosion of channel banks (Foster, 2008). Also based on some limited geochemical fingerprinting, Berry (2005) suggested that bank erosion in the upper catchment tributaries, as well as erosion of agricultural land, may make a major contribution to sediment delivery to the lower catchment.

In headwater stream channels, an increase in organic matter is often associated with an increase in invertebrate communities as a result of the improved availability of food. However, flocculation with mineral sediment can cause organic matter to accumulate on the stream bed, increasing the oxygen demand through microbial activity at the expense of aquatic invertebrates. Nicholls (2015) reported organic matter in sediment in the range of 3.17–8.66% which is equal to and generally higher than the average value we have recorded in the upper 10 cm of fields in the catchments (3.17 ± 0.32%). As well as the influence of mineral sediment, this elevated organic matter concentration can be explained by contributions from additional sources such as riparian habitats and woodland with higher soil organic matter. Alsolmy (2016) attributed the higher abundance of dipterans (mainly chironomids) and oligochaetes in local arable ponds and streams to the organic-rich sediments present there.

Fine particles act as a vector for pollutants such as phosphorus and some pesticides. Ockenden et al. (2014) explored the relationship between sediment particle diameter and nutrient concentrations across a range of soil types, including the clay soils at Loddington and found correlations for total phosphorus, total nitrogen, and carbon. For each element, particles of less than 10 μm were associated with the highest concentrations. The propensity for phosphorus to adsorb to sediment particles means that erosion of topsoil is a major pathway for this nutrient to enter watercourses. Phosphorus in water can be separated into this particulate form and total dissolved phosphorus which in turn comprises soluble reactive phosphorus (SRP; orthophosphate) and more complex dissolved organic phosphorus (DOP). SRP is the most readily available form and the one which has the greatest impact on aquatic or terrestrial plants.

Phosphorus and aquatic ecology

Wasiak (2009) explored the relationship between phosphorus concentration and diatom and aquatic invertebrate communities in small headwater streams at and around Loddington, comparing a mainly arable catchment with a low input pasture one. He collected additional comparable data from similar streams in the Herefordshire Wye and Hampshire Avon river basins and incorporated them into the same analysis. The overall diversity of macroinvertebrates was generally higher in the sites with low P concentrations. Both Simpsons and Margalef diversity indices for invertebrates were negatively correlated with suspended sediment, particulate phosphorus, SRP and total phosphorus.

Consistent with these findings for diversity, the ratio of macroinvertebrate biomass (dry weight) to the number of taxa was higher in the low phosphorus control sites than in streams with high P concentrations. In other words, the biomass was distributed among a lower number of taxa in the streams with high P concentrations than in the less impacted ones. Invertebrate biomass varied considerably up to SRP concentrations of 110 μg/L, but above this concentration, biomass was consistently high. Invertebrate biomass was therefore highest at sites with high P concentrations while the number of taxa found was lowest at these sites. High nutrient concentrations therefore support pollution tolerant taxa at the same time as reducing species diversity.

The site in the Eye Brook headwaters was the closest of all the English sites studied to a natural state, with phosphorus concentrations often below the level of laboratory detection (7 μg/L SRP). Spring and autumn chlorophyll *a* biomass for these Eye Brook headwaters were also the lowest recorded across the whole project. This site also supported the highest benthic diatom diversity and the highest macroinvertebrate diversity. The latter was dominated by the functional group 'shredders', and macroinvertebrate predators represented more than 5% of the invertebrate community. In contrast, at the more impacted sites, with SRP concentrations ranging between 30 and 120 μg/L, predators represented less than 0.1% of the invertebrate community.

The contribution of shredder invertebrates (mainly freshwater shrimps: Gammaridae) decreased, and collectors represented by dipterans (mainly Chironomidae) took their place as P concentrations increased. Chironomids were mainly filter feeders, obtaining food from the water column. High microbial respiration was recorded at these sites, indicating high microbial biomass, but decomposition of organic matter was slower. This may have been because of the reduced shredding activity associated with low numbers of *Gammarus*.

Alsolmy (2016) found that macroinvertebrate communities differed between arable and pasture streams, but this was not the case for ponds in arable and grass catchments. Arable streams had significantly higher numbers of oligochaetes and lower numbers of Plecoptera than pasture streams, and in terms of functional feeding groups, shredders and scrapers were more abundant in pasture than arable streams. Leaf litter processing did not differ between arable and grass, but leaf litter breakdown rates increased with the abundance of shredders. In contrast to the results for macroinvertebrates, diatom species richness and diversity were lower in arable ponds than in pasture ponds, but there was no such difference between land uses for streams.

In the streams studied by Wasiak (2009), the Margalef diversity index of diatoms was negatively correlated with SRP (Defra, 2008). A relatively rapid decrease in diatom cell count occurred below a concentration of around 100 µg/L. The contribution of non-diatom algae to community composition increased with phosphorus concentration and at sites with SRP in the range of 30–120 µg/L, it was close to 90%. For chlorophyll *a* biomass, a significant negative correlation was found with SRP, particulate phosphorus and total dissolved nitrogen. Above 100 µg/L SRP chlorophyll *a* biomass increased rapidly, decreasing again above a concentration of 120 µg/L as limiting factors other than phosphorus came into play.

Because of their location in dense riparian corridors often flanked by high banks and trees, small headwater streams had reduced exposure to light and this was the second most important variable influencing algal biomass at these higher concentrations. Much of the ecological change in diversity and biomass therefore takes place below a threshold of around 100 µm/L SRP. Statutory targets for 'Good' water quality status in local streams range from 60 to 90 µm/L SRP and so fall within the range in which our data suggest most ecological change is occurring.

Jarvie et al. (2010), working in the same small headwater streams across the UK, identified strong seasonality of SRP, with high concentrations, sometimes exceeding 1,000 µg P /L occurring between July and September. Lower concentrations (typically <200 µg/L) occurred between December and February. There was also strong flow-related seasonality in NO_3 concentrations, peaking at >10 mg N /L between November and March, and falling below detection levels (<0.1 mg N/L) between July and September.

These seasonal changes can be attributed to the nutrient contributions from domestic sources in the form of septic tanks associated with rural

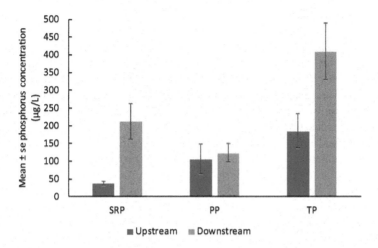

Figure 6.2 Concentrations of soluble reactive phosphorus (SRP), particulate phosphorus (PP) and total phosphorus (TP) upstream and downstream of a septic tank discharge into a small stream at Loddington (drawn from data in Withers et al., 2011)

houses. Although septic tanks have associated soakaways which are intended to allow water discharged from the tanks to enter the subsoil, these are ineffective on clays soils, allowing high nutrient status water to enter directly into watercourses.

Withers et al. (2011) investigated the role of these small domestic sources of phosphorus in the same stream. Effluent discharges were highly concentrated in the range of nutrients that are typically associated with inorganic products of anaerobic digestion of organic material in septic tanks and associated detergent ingredients; SRP, NH_4N, Na, Cl, B and Mn. SRP and NH_4N were particularly well correlated with the detergent marker Na. The high ammonium concentrations are indicative of very high levels of organic pollution which can have severe consequences for aquatic organisms through the toxic effects of ammonia and oxygen depletion. Fish are particularly sensitive to ammonia toxicity and oxygen stress. These continuous (point) sources dominated P concentrations for much of the year during summer base flow when the ecological impact was greatest (Figure 6.2).

The annual Total Phosphorus (TP) load attributed to these base flow point sources was 0.82 kg per year at the Eye Brook headwater site, but considerably higher, at 25.3 kg per year at one of the Loddington sites (Jarvie et al., 2010). Despite the importance of these small domestic sources of nutrients in terms of ecological impact during the summer, flow-dependent (diffuse) sources of TP from the catchment as a whole contributed more than 95% of annual TP loads exported from the catchment. Taking account of catchment size, these diffuse sources amounted to around 500 g/ha/yr.

This compares with the estimate of 70–380 g/ha/yr derived from plot experiments at Loddington (Deasy et al., 2008, 2010; see Chapter 3). A likely explanation for these lower values is that the direct measurement of runoff from arable land was based on consistently planar experimental plots, while catchment scale estimates include sources associated with undulating topography where surface runoff is concentrated, flow is increased and along with it, the concentration of suspended sediment and phosphorus. While domestic point sources of nutrients have the greatest ecological impact within catchments, diffuse sources associated with erosion are responsible for the greatest export from the catchment, and ultimately to coastal waters.

Palmer-Felgate et al. (2010) examined nutrient release into water from bed sediment under anoxic conditions at a septic tank discharge site where sewage fungus (a development of filamentous bacteria, fungi and periphyton) was associated with discharged organic matter, increasing oxygen demand. She compared this sampling point with one 12 metres upstream and another 8 metres downstream of the discharge pipe and recorded sediment concentrations of TP, TN and TC that were lowest upstream, highest at the discharge site, and intermediate downstream. There were extremely high concentrations associated with this domestic source. At the discharge site, SRP and ammonium had a peak just above the interface between fungus and sediment, SRP concentrations reaching 13,627 µg/L compared to 4,273 µg/L at the outlet of the discharge pipe itself. At the downstream site, SRP in the sediment at 8.5 cm depth reached 10,529 µg/L compared to 619 µg/L in sediment at the upstream site.

These findings demonstrate that effluent discharge into streams with a low dilution capacity can lead to increased sediment phosphorus concentrations further downstream. This phosphorus can subsequently be released into the sediment pore waters where it can act as a direct nutrient source for macrophytes and benthic algae. Even without extreme anoxic conditions such as those created by the sewage fungus, SRP can be remobilised from the sediment into the water. We can therefore expect phosphorus concentrations in headwater streams to increase in response to ongoing climate change as a result of reduced dilution of these domestic sources, and remobilisation from bed sediment under anoxic conditions.

The potential role of sediment traps

Chapter 3 described the process of surface runoff, soil erosion and transport of nutrients to water, and some means of mitigating this. However, there has also been interest in mitigation measures outside the cropped area. One potential approach to reducing the agricultural contribution of sediment and associated P to water is the creation of sediment traps at field edges to capture eroded soil before it enters watercourses. We have investigated the potential of small field corner ponds to capture eroded sediment and reduce ecological impacts in water at Loddington (Ockenden et al., 2014). The ponds were of three designs: a single shallow cell to receive surface runoff, a pair of shallow cells, and a single deep cell connected to a single shallow cell (Figure 6.3). As well as being created on the clay soils at Loddington, the same designs were adopted on sand and silt soils in northwest England.

Sediment accumulation in the ponds varied considerably between the three soil types, being highest on the sand soil (0.8 t/ha/yr), intermediate on silt soils (0.3 t/ha/yr), and lowest on the clay soils at Loddington (0.04 t/ha/yr) (Figure 6.4). Accumulation of total phosphorus, total nitrogen and carbon followed the same pattern, being consistently lowest at Loddington. Although phosphorus accumulation in the ponds was higher on sand and silt soils than on the clay soils at Loddington, even at those sites this represented less than 1% of the P applied as fertiliser to the area of land draining into the pond. The implication of this is that the sediment accumulated in field corner pond systems has little value in terms of nutrients that can be returned to the fields from which they came.

Although lower rainfall at Loddington contributed to lower soil erosion and subsequent accumulation in the ponds, sediment capture was driven mainly by differences in erodibility and transport of the three soil types. The highest proportion of eroded soil was captured on sandy soils, whereas fine particulate matter from eroded clay fields at Loddington was held in

---

Box 6.1 Sediment traps

Sediment traps are ponds that are constructed specifically to capture sediment moving towards watercourses via surface runoff. They are less effective at capturing material transported in field drains because most of this is held in suspension. Similarly, especially on clay soils, only the coarser material will be captured. Sediment traps therefore need to be sited carefully according to soil type and known surface runoff pathways. The lowest corner of a field where a ditch can be diverted into the sediment trap is an ideal location.

The precise design needs to be flexible to accommodate constraints or opportunities at individual sites. The larger the sediment trap, the longer the residence time and consequent sedimentation, but the larger the area of ground lost to production. Creating a two-level process involving a deep receiving pond followed by a shallower one that is planted with reeds or other robust vegetation provides an opportunity for coarser material to be trapped at the first stage and for finer material to settle out as water passes through the vegetation. However, creating an initial deep pond results in a larger amount of spoil which will need to be incorporated into the pond design (resulting in additional loss of land from production) or transported away from the site (adding to the construction costs).

Figure 6.3 Sediment trap created for research showing deep receiving pool in the foreground and shallower pool colonised by common reed *Phragmites australis* at discharge point in the background

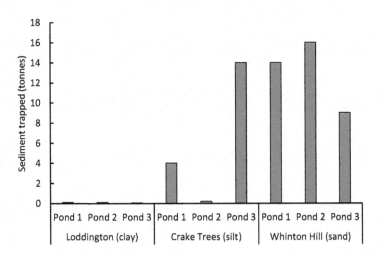

Figure 6.4 Sediment trapped in ponds on sand, silt and clay soils in 2010/2011. (Drawn from data in Ockenden et al., 2014)

suspension through storm events and beyond, and consequently passed through the pond systems and into receiving watercourses.

The implications of this research are that field corner pond systems designed as sediment traps are ineffective on clay soils unless they are receiving coarse material associated with surface runoff. The emphasis for management on clay soils needs to be minimising the loss of soil from fields in the first place, a topic covered in Chapter 3. Arguably, for sand soils, the conclusion could be the same, as the large amounts of sediment captured by the pond systems indicate a substantial problem in terms of soil loss from fields. However, our results suggest that on sand and silt soils, these pond systems offer a potential supplementary course of action for reducing the impact of soil erosion on aquatic systems.

Catchment scale research

Background

Catchment processes have been investigated at the wider landscape scale since 2010 in the Water Friendly Farming project which incorporates the small Eye Brook headwater stream studied by Withers et al. (2011) and Wasiak (2009). The Water Friendly Farming project study area lies within the Eye Brook and Stonton Brook sub-catchments of the River Welland and the Barkby catchment in the River Soar in Leicestershire (Biggs et al., 2016, Figure 6.5). Each headwater catchment is around 10 km^2 in area. The catchments are directly adjacent and have very similar geologies, topographies and land uses. This is a region of low rolling hills (95–221 m asl) with

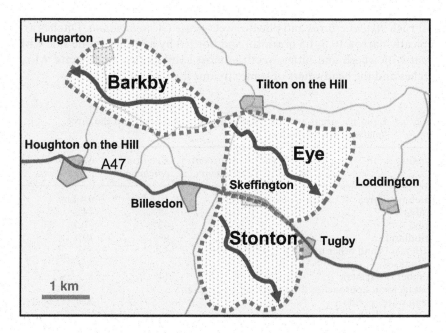

Figure 6.5 Water Friendly Farming study area showing the two Eye Brook and Stonton Brook headwater 'treatment' catchments of the Welland and the 'control' Barkby Brook catchment of the Soar headwaters

high ground areas (including most headwaters) dominated by Pleistocene fluvio-glacial sands, gravels and clays. Valley sides are predominantly Middle and Lower Jurassic mudstones and siltstones with beds of ironstone. In valley bottoms, the Jurassic strata are overlain by recent deposits of alluvium or colluvium. Cultivated land falls into two of the most extensive of Britain's agricultural land classes which make up 35% of the cultivated land in Great Britain (Brown et al., 2006).

Agriculture in all three catchments is mixed farming, divided between arable land mainly under winter wheat with break crops such as rape, field beans, barley or oats, and grassland used for beef cattle and sheep or cut for hay or silage. Farm sizes vary considerably across the study area, and there is also a range of tenure arrangements including owner-occupied, short- and long-term tenancies, contracting and share farming agreements. Table 6.1 shows the proportion of land cover types in each catchment.

The study area has three main waterbody types based on the definitions given in Williams et al. (2004): streams, ditches and ponds. Streams in all catchments are relatively small with a maximum width of c.3 m. The lower reaches of all streams are typically shaded by riparian trees with their margins supporting shade-tolerant plants such as great willowherb *Epilobium hirsutum*, bittersweet *Solanum dulcamara*, angelica *Angelica sylvestris* and meadowseet *Filipendula ulmaria*. In the upper reaches of all catchments, some stream lengths are more open and bordered by fenced or unfenced pasture often with the channels supporting brooklime *Veronica beccabunga*, soft rush *Juncus effusus* and plicate sweet-grass *Glyceria notata*. Ditches are typically narrow (0.5–1.5 m width) and borded by a hedge on one side with great willowherb and bittersweet the most common wetland plants. Most ditches and the headwaters of some streams flow seasonally.

Table 6.1 Land-use areas and management practices adopted in the three headwater catchments

Landuse	Barkby control catchment	Eye Brook catchment	Stonton catchment
Catchment area	9.6 km^2	10.6 km^2	9.4 km^2
Arable	37%	45%	44%
Grass	52%	42%	41%
Woodland	7%	9%	10%
Settlements and other minor landuses	4%	4%	5%
Number of farmers	7	14	8
Management approaches discussed in this chapter			
Clean water ponds			X
Permeable dams		X	
Sediment traps		X	X
Soil management		X	X

Ponds show considerable between-waterbody differences in terms of their shading and seasonality. In pasture areas, most ponds are fenced. In addition to supporting the plant species that are widespread in streams and ditches, ponds commonly support grasses and rushes such as hard rush *Juncus inflexus*, tufted hair grass *Deschampsia cespitosa*, floating sweet-grass *Glyceria fluitans* and branched bur-reed *Sparganium erectum*. They are also the only habitat to consistently support floating and submerged leaved plants, the most widespread species being rigid hornwort *Ceratophyllum demersum*, small pondweed *Potamogeton berchtoldii* and common duckweed *Lemna minor*.

A comprehensive census of aquatic plants in wetland habitats (ponds, ditches and streams) was carried out each year across the three headwater catchments (Williams et al., 2020). Whereas a representative survey would be necessary for most terrestrial habitats, the restricted distribution and coverage of these small wetland habitats make a census of entire plant populations feasible.

Total phosphorus, total nitrogen, suspended sediment and flow have been monitored at the base of each of the three headwater catchments since 2012. In particular, we have explored issues associated with phosphorus in water, building on the research described above. Concentrations of some pesticides were also monitored during the autumn and winter for four consecutive years, and we explored issues associated with reducing pesticide concentrations in water.

Stream water depth was monitored at the base of the headwater catchments every 15 minutes and then converted into stream flow (m^3/s) using a flow rating curve generated for the catchment. The Soil and Water Assessment Tool (SWAT; Arnold et al., 1998) was used to simulate stream flow. SWAT is a physically based hydrology and water quality model, designed to estimate impacts of land management practices on water quality in complex watersheds. SWAT divides the catchment area into sub-catchments and each of them is further divided into hydrological response units which are defined as areas of land with the same soil, land use, slope and management which are assumed to behave similarly in the model (Neitsch et al., 2005). We used Environment Agency estimates of sewage treatment works discharges, estimates of livestock grazing, and farmers' fertiliser and pesticide input data wherever possible, as input to the model for analysis of phosphorus and pesticide concentrations in water.

From 2014 to 2018, a number of mitigation measures were introduced in the two treatment catchments (Table 6.1). In the first post-baseline phase, earth dams in ditches, and on-line and off-line detention ponds to capture sediment before it entered the mainstream were created across both catchments. In the Stonton catchment, 20 clean water ponds were created in micro-catchments that were identified as being unimpacted by sources of pollution, and numerous small woody debris dams were introduced into the smallest streams. In the second phase, the focus was more on flood risk management. In this case, 27 permeable timber dams were built in 2017 and 2018 in streams and large ditches to hold water back during peak flow in order to reduce flood risk downstream. Throughout the project, the third catchment (Barkby) remained as a control. No mitigation measures were installed here although, as in the Eye Brook and Stonton, this catchment had a normal level of agri-environment scheme protection such as riparian buffer strips. The results of these various elements of our research programme are discussed below.

Phosphorus in water

Total phosphorus concentrations at the base of the three headwater catchments varied considerably through the year. Concentrations were generally highest at baseflow during the summer and early autumn in line with the findings from the small catchment at Loddington described earlier (Jarvie et al., 2010), but some high concentrations were also associated with high flows during storm events (Biggs et al., 2016). A total of 91 grab samples collected from four tributaries in the Eye Brook catchment, and at the base of the catchment, between January 2012 and January 2016 revealed that the tributary in which a sewage treatment works was located had consistently higher P concentrations than the other tributaries (Figure 6.6).

Modelling source apportionment of P also revealed a high contribution from the sewage treatment works, contributing 24–45% of the total load. However, the background contribution comprising mainly runoff from agricultural land was estimated to be 23–57%. Whereas in the small catchment at Loddington, septic tanks made a major contribution, in the larger headwater catchment the septic tank contribution was only 0.13–0.56% and the domestic sources were derived almost exclusively from the sewage treatment

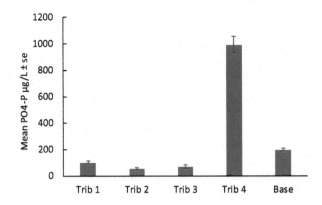

Figure 6.6 Phosphate concentrations at the base of four agricultural tributaries in the Eye Brook headwater, and at the base of the headwater study catchment. Tributary four contains sewage treatment works

works. Livestock manure and fertiliser applied to arable crops made up the balance. The four-year average background contribution from diffuse sources was 41%, considerably lower than the 95% estimated for the small Loddington catchment by Jarvie et al. (2010).

We used four land-use change scenarios to explore the potential for reducing P concentrations in the Eye Brook. One was an increase in buffer strip width against all watercourses to 20 m. Another was conversion of arable land to pasture to achieve the proportion of pasture that was present in the 1930s. An 'afforestation' scenario converted all agricultural land to woodland, and the final scenario was the removal of the sewage treatment works. The results are presented in Figure 6.7. Reductions in P for conversion of arable land to pasture were similar to those achieved by creating 20 m wide buffer strips, although the model may exaggerate the potential of the latter as it does not account for the concentrated flow associated with undulating topography. Removal of the sewage treatment works was equivalent to complete afforestation of the agricultural area.

Given that movement of P to water from agricultural land is primarily associated with movement of sediment to watercourses and losses from manures, we also explored the potential impact of afforesting agricultural land on reduction in the annual suspended sediment load. We compared this with 20 m wide buffer strips and the conversion of all arable land to direct drilling. Afforestation was associated with a 74% reduction in sediment load, while wide buffer strips and direct drilling were 26% and 33% respectively (Figure 6.8).

Further analysis of river phosphorus concentration data from sites across the wider upper Welland catchment (RePhoKUs project, 2021) provided more evidence for the dilution of SRP from domestic sources (sewage treatment works and septic tanks) with increasing flow and showed that mobilisation of P from agricultural land occurred at higher flow rates exceeding

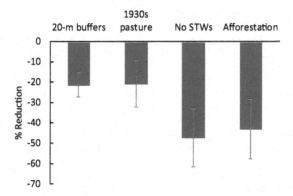

Figure 6.7 Percentage changes in phosphorus concentrations in water in response to modelled land-use change scenarios (Brown & Velez, 2021)

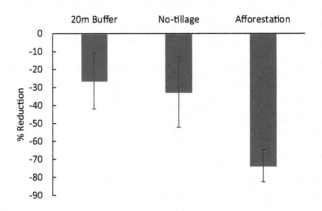

Figure 6.8 Percentage changes in suspended sediment concentrations in water in response to modelled land-use change scenarios (Brown & Velez, 2021)

around 5 L/Km²/s (3–7 L/Km²/s). SRP concentrations at high flow varied considerably between sites, probably related to the type of diffuse sources present. High-flow TP losses were more consistent and were not higher in autumn and spring when P fertiliser is applied, suggesting that soil erosion is the largest source, rather than direct loss of applied fertiliser.

An annual phosphorus balance was carried out for the food system in the Welland river basin, based on data from a wide range of sources (Re-PhoKUs project, 2021). Fertiliser was the largest import of P into the catchment, but P offtake in crops and grass exceeded manure and fertiliser inputs. This negative balance will effectively draw down reserves of legacy P stored in the soil by approximately 4 kg/ha/year. Compared to a UK P use efficiency of 65%, the equivalent figure for the Welland was 135%. Source apportionment analysis reveals that, alongside this, there was still a 39% loss of P from agricultural land to water, with 61% from sewage treatment works and septic tanks (RePhoKUs project, 2021). This contrasts with the food production system in the Wye river basin where there was a substantial positive phosphorus balance, largely associated with the livestock systems present in the area. Even in the Welland, an estimated phosphorus export rate of 0.4–0.7 kg/ha/year results in substantial impacts on aquatic ecology as discussed earlier in this chapter despite high P use efficiency in terms of food production.

Application of the DESPRAL soil dispersion test (Withers et al., 2007) to soil samples from the two river basins reveals higher phosphorus concentrations in runoff from soils in the Wye than the Welland river basin. In both areas, there is therefore potential for more efficient management of soil phosphorus to improve uptake by crops and reduce losses to water, as discussed in Chapter 3.

Pesticides in water

Propyzamide is an important herbicide in the very limited range of plant protection products that can be used to combat black-grass *Alopecurus myosuroides*, a highly competitive arable weed. It is applied mainly at the oilseed rape stage of the rotation, but also in field beans. However, water companies find it difficult to remove this herbicide from the drinking water supply and the concentration can only be kept below the statutory 0.1 µg/L limit required at water treatment plants by dilution from other sources of supply.

Black-grass thrives in compacted waterlogged soils which are also most susceptible to erosion, and constrain crop growth and yields. Propyzamide is applied to soils with relatively high moisture content and it moves to water both in solution and adsorbed to soil particles. Reducing soil erosion therefore helps to reduce propyzamide concentrations in water as well as achieve other water-related objectives. There are therefore commonalities between farmer objectives for productive and profitable crop production, water company objectives for reducing pesticide concentrations in water, and environmental objectives for the conservation of aquatic wildlife. Because sediment reduces the water storage capacity of drainage channels downstream, there are also implications for flood risk management. The herbicide is at the centre of a web of interacting issues.

We used four years of herbicide concentration data from our study catchments in the Water Friendly Farming project, together with rainfall, stream flow, and crop area data, to model the potential for reducing the concentration of propyzamide to below the 0.1 µg/L limit at the base of the Stonton headwater catchment (Stoate et al., 2017; Villamizar et al., 2020). The modelling identified rainfall and oilseed rape crop area as the main drivers of propyzamide concentrations in the stream. To keep the herbicide concentration below the statutory limit, it would be necessary to restrict the oilseed rape area to within 2–3% of the catchment area. Additional or complementary mitigation options include extending the arable rotation with vigorous hybrid barley which can compete against black-grass (reducing the density of the weed) and providing an alternative break crop. Increasing the width of riparian buffer strips to 20 metres reduced the concentration by 25–70%. Reducing tillage intensity reduced sediment and associated herbicide loss to water but the modelling suggested only a 5–16% reduction in propyzamide concentration because of the improved connectivity between the soil surface and field drains created by earthworms in undisturbed soil (Brown & Beinum, 2009). Improved knowledge of soil compaction within fields could potentially also guide management that would reduce runoff, and availability of local soil moisture data could improve the timing and efficacy of mitigation practices such as subsoiling and mole ploughing as well as application of the herbicide itself.

We put these various options to the farmers in the Stonton headwater. In some years, the oilseed rape area was around 5% of the catchment, but in others it had been up to 30%, depending on which part of each farm's crop rotation was within the catchment boundary in any one year. Restricting the area to 2–3% of the catchment in order to keep within the statutory limit would impose too heavily on the economics of farm businesses and create tensions between neighbouring independent farms. Multiple tenure arrangements between neighbouring farms complicate the issue, and expansion of the area managed by contractors restricts long-term planning and makes timely operations more difficult. The farmers also expressed concern about the economic risks associated with adopting a direct drilling approach to soil management on clay soils.

Introducing hybrid barley into the rotation was well received by the participating farmers as a means of extending the rotation and controlling black-grass. While 20-metre wide buffer strips were considered to be a major constraint in fields that are often relatively small, the farmers were accepting of the principle of buffer strips and had adopted narrower strips (normally of up to 6 m width) for many years. There was also interest in improving understanding of compaction with a view to carefully targeted management that would reduce erosion and improve crop performance. Access to local soil moisture data was also considered to be potentially useful to guide appropriate management of both pesticides and soils. While farmers were reluctant to collaborate to meet pesticide concentration targets, they agreed that there was an important coordination role for local trusted advisors, supporting the findings of Morris and Jarratt (2016) for the wider Welland river basin and other areas.

The farmers were keen to emphasise that the 0.1 µg/L was arbitrary and not based on risk to human health or environmental impacts and that a higher limit would be both more meaningful and achievable. In fact, the 0.1 µg/L limit applies to water supplied at the tap following abstraction from the river many miles down the catchment. As there will have been dilution and degradation of the herbicide over that length of river, a limit of 0.5 µg/L might be more appropriate in headwaters. However, this would require constraining the rape area to 8–10% of the catchment area which remains well below the area of 30% actually occurring in some years and the farmers could not see how this would be managed at the catchment scale.

While the focus in this study was on a specific herbicide to meet statutory requirements for drinking water supply, because of the process by which propyzamide reaches watercourses, the lessons learnt have implications for catchment management objectives for nutrient concentrations in water, flood risk management and landscape scale aquatic ecology. What has become clear in this study is the need for a broader focus than simply on the use of the herbicide itself to encompass wider aspects associated with the control of black-grass including soil management and crop rotation.

As such, this integrated approach is relevant to the use of other plant protection products and the control of their target pests, weeds and diseases. There is further discussion of this issue in Chapter 8.

Clean water ponds

Aquatic invertebrate data collected from ponds at and around Loddington contributed to a pan-European study of the relationship between aquatic biodiversity and land use across Belgium, Denmark, Hungary, Germany, Poland and Slovenia, as well as the UK (De Bie et al., 2010). The study highlighted the strong negative impact of arable land use on aquatic biodiversity, especially within 200 m of the surveyed ponds. The authors caution that, as ponds in arable landscapes tend be present at lower densities than in other land uses, this impact may be partially explained by isolation and reduced capacity for dispersal of aquatic organisms. However, from our own research described above, and from similar findings elsewhere, we know that there are substantial negative impacts of nutrients and other pollutants from arable land use on aquatic species. This raises the possibility that identifying micro-catchments that are not affected by cropping and creating new wetland habitats in those areas could contribute to aquatic species diversity at the landscape scale. We explored this option by creating new ponds in the Stonton catchment.

The existing ponds across the Water Friendly Farming study area were a key habitat in the agricultural catchments that we studied and supported the greatest number of wetland plant species, the most uncommon species and the highest proportion of unique taxa, with 40% of the species recorded only occurring in ponds (Figure 6.9). We know from Willams et al. (2004) that these results for plants are also likely to be reflected in aquatic invertebrate communities.

Taking all three habitats together, 106 aquatic plant species were recorded across the landscape over the initial nine years of the study, with slight year-to-year variations. The overall temporal trend was downwards, with a 10% decline in the number of species over the nine years. The rate of decline differed between habitats, being greatest in streams (1.2% pa), and intermediate in ditches (0.7% pa), with no significant change in ponds. In contrast, when considered separately, rare species declined by 22% across the study area over the nine years, with around 70% of this loss being attributed to submerged aquatic plant species in ponds. For ponds at least, national Countryside Survey results broadly parallel our study findings, showing that in the English lowlands, pond alpha plant species richness declined by around 20% (1.8% pa) between 1996 and 2007 (Williams et al., 2010).

The wetland flora in our survey area remain vulnerable to further loss. During the 2010–2018 survey period we recorded the extinction of some species from all catchments and waterbody types through habitat change and destruction including culverting of springs, planting of trees in a small fen, and cessation of grazing along waterbody margins.

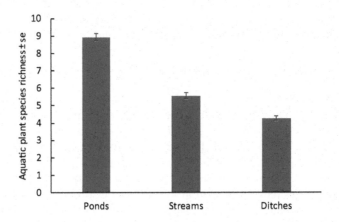

Figure 6.9 Aquatic plant species richness in ponds, ditches and streams across the Water Friendly Farming study area (Penny Williams, Freshwater Habitats Trust data)

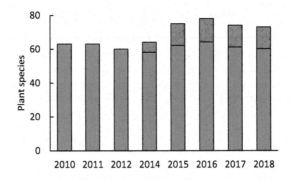

Figure 6.10 Aquatic plant species richness in small water bodies, 2010–2018, with new species associated with the creation of clean water ponds in 2013 represented in orange (Penny Williams, Freshwater Habitats Trust data)

After the two-year baseline period, we created twenty clean water ponds (Figure 6.11) specifically for wildlife in the Stonton headwater catchment. We selected sites carefully so that the new ponds were in low input pasture or open areas of woodland where runoff into them was not affected by domestic or agricultural sources of nutrients; none were connected to inflowing streams and ditches. These new ponds occupied 0.24-ha of the 9.4 km^2 catchment. Creation of these clean water ponds more than compensated for recent losses and, after five years, had increased Stonton catchment aquatic plant species richness by 26%, and the number of rare plant species associated with increased by 181% (Figure 6.10; Williams et al., 2020).

Box 6.2 Clean water ponds

A fundamental requirement of ponds that are being created to benefit aquatic wildlife is that they are not impacted by adjacent human activities. In most cases, this will be agricultural land use. Selecting sites in micro-catchments where there is no such impact is therefore essential and might comprise low input grassland or open woodland where there is no surface runoff or field drains from more intensively managed agricultural land. Relying on runoff from the micro-catchment, rather than a stream or spring ensures that the water quality is not compromised by human activities elsewhere. The site must also have an impermeable substrate to avoid the need for a clay or synthetic lining. Another important consideration is that ponds should not be created where there is an existing species-rich habitat such as meadow or marsh, or archaeological features.

Use of a 360° tracked excavator to create the pond optimises versatility and provides an opportunity to compact pond banks where necessary to minimise water infiltration. Where there is sufficient space, wildlife value can be maximised by creating a complex topography with broad undulating drawdown zones, underwater bars, and shallow slopes, as well as more permanent water of up to 1.5 m depth. This habitat complexity can be increased by creating a series of separate ponds in the same location. As well as the spatial variation in water depth, seasonal changes in the water level also benefit wildlife.

Figure 6.11 Clean water pond in the Stonton headwater catchment

(Continued)

Topsoil should not be spread around the edge of the pond or anywhere where sediment and nutrients from it can be washed back into the pond. Similarly, other spoil from the excavation needs to be placed away from the pond where it has no negative impact on wildlife habitats and could even be used to create new ones. Spoil can be removed from the site, but this can add considerably to the cost.

Plants colonise ponds very quickly, so it is not necessary, or desirable, to introduce plants from other sites. Low-level grazing is beneficial to plant and associated invertebrate communities and fencing can be used to control this but should not be used to exclude grazing livestock completely unless the stocking density is very high.

Pond creation can be considered as an ongoing process, with modifications being made in subsequent years as the water levels become better understood and the vegetation develops.

More detailed guidance on pond design and creation is available at www.freshwaterhabitats.org.uk.

The high proportion of rare species that appeared in the catchment's newly created clean water ponds is in line with findings from other studies. Williams et al. (1998) showed that 6–12-year-old ponds supported more uncommon plant species than older ponds in lowland Britain, while Fleury and Perrin (2004) found that populations of threatened temporary pond plants were greatest in the first 2–3 years after ponds were created.

Rare species are often considered to be poor competitors and the bare substrates of new ponds may provide a competitor-free zone in the first few years after creation. Recently excavated waterbodies also tend to have relatively nutrient-poor substrates which may directly benefit groups including charophytes that have been shown to thrive below relatively low nutrient thresholds (Lambert & Davy, 2011). Alternatively, pond creation may have exposed seed banks, allowing rare plants that were no longer present in the existing flora to germinate (Nishihiro et al., 2006).

In our study, it seems likely that more than one factor was involved. A proportion of the nationally and regionally rare species that we recorded (e.g. marsh arrow-grass *Triglochin palustris*, blunt-flowered rush *Juncus subnodulosus*, bristle club-rush *Isolepis setacea*) appeared only in ponds created in an area of secondary woodland partly planted on a former fen. Marsh arrow-grass, at least, had previously been recorded from the fen, and although pre-excavation surveys did not find it in the area where the ponds were created, it seems likely that this species germinated from a pre-existing seed bank. Other taxa, including marestail *Hippuris vulgaris* and *Chara* species, appeared in new ponds that had been created in isolated dry ground areas where colonising species can only have arrived through wind or bird transported propagules.

The submerged aquatic lesser pondweed *Potamogeton pusillus*, was not present in the Stonton catchment before the new measures were introduced. It rapidly colonised a number of the new clean water ponds and was subsequently recorded in first one, and then a second pre-existing pond, around 1 km away. Such incidents provide a strong indication that in some cases, the introduction of new ponds into the landscape has improved connectivity, which ultimately helped to support greater alpha and gamma richness in the catchment's pre-existing waterbodies. That the introduction of new ponds reduced the average distance between ponds from 255 m to 92 m will have facilitated this process.

Fencing of clean water ponds and streams influences the plant and invertebrate species present. A widespread practice to prevent damage to ponds and streams is the fencing out of livestock. From an agricultural perspective, this can be an important means of isolating livestock from sources of liver fluke; this keeps animals clean, and where streams are small and easily crossed by animals, this also helps to sub-divide pastures so that they can be more efficiently grazed. Fencing livestock away from water prevents deposition of dung and urine and has been demonstrated to have rapid benefits to stream ecology, not just by reducing bankside soil erosion and deposition of nutrients into water, but by physical modification of the channel. The establishment of dense tall vegetation along the bank results in lack of grazing, which in turn constrains the channel width, increases water flow, reduces streambed sedimentation and increases the heterogeneity of habitats within the channel.

However, some plant and associated invertebrate species are dependent on the grazing of pond or stream bankside vegetation and the loss of livestock can result in loss of biodiversity from this habitat. Monitoring of plants along the banks of one fen stream within the 'control' catchment

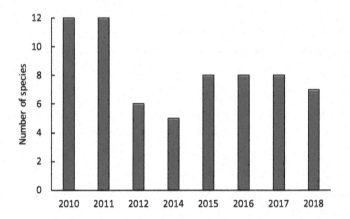

Figure 6.12 Plant species richness before and after fencing of a headwater stream in 2011 (Penny Williams, Freshwater Habitats Trust data)

of the Water Friendly Farming project revealed a reduction in plant species richness following fencing. The stream was fenced by the landowner in 2010 and half the plant species present in that year were lost within two years (Figure 6.12). While riparian fencing can often have considerable ecological benefits, botanically diverse habitats can be affected negatively by fencing. Where there is existing public access, fencing prevents people from having physical contact with the water and wildlife associated with it, creating a disconnect between the local community and their environment.

Flood risk management

There is widespread and increasing interest in Natural Flood Management such as afforestation, off-line flood storage ponds and leaky timber dams in headwaters to complement traditional engineered flood defences in flood risk areas. Although leaky timber dams are being widely deployed, there has been little evaluation of their efficacy. To address this, we introduced 27 such in-channel structures, of varying sizes, into the Eye Brook headwater to evaluate their potential for attenuation of downstream flood peaks.

Locations in ditches and streams were selected carefully to optimise water storage behind them while avoiding impeding flow from field drain outlets and waterlogging adjacent productive land. The dams were of simple construction and initially created with mainly local materials by a local contractor. Standard timber (mainly larch) cordwood was the main component, held in place with tanalised fence posts and steel cable. A tree trunk formed the base of each dam to ensure that winter base flow was not impeded. This design worked well in minor watercourses but was not sufficient to withstand high flows in the main channel. These larger dams were therefore rebuilt using 8 m lengths of tree trunk, spanning the whole channel width, to remove weak points associated with shorter lengths of cordwood.

Field measurements of channel profiles along different streams within the catchment were taken to validate and correct LIDAR's Digital Evaluation Model to minimise errors due to limitations in its spatial resolution and issues with penetration capabilities through water and vegetation. The actual permeability of leaky barriers was studied by placing time-lapse cameras next to several barriers to record water depth across peak flow events. Gauge boards were placed upstream and downstream of some of the barriers to record water depth. These data were used to adjust permeability factors in the model to obtain the closest match to the behaviour of water depth curves between the model and the observed recordings for the dams. The leaky barriers added about 17,700 m^3 of water storage (1,621 m^3/km^2) into the headwater catchment.

The model predicted significant reductions in peak flows at the base of the headwater catchment for the most frequent design flood events (e.g. 14.5%

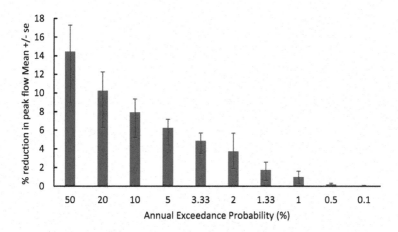

Figure 6.13 Modelled reduction in the base of catchment peak flow associated with permeable dams for a range of annual exceedance probabilities. 10 = 1 in 10-year storm events, 1 = 10 in a 100-year events, and 0.1 = 1 in a 1,000-year events (Brown & Velez, 2021)

reduction for the 1-in-2 year event) with a reduction in barrier effectiveness as the intensity of the event increased. For example, we estimated 1.0% and 0.1% reductions in peak flow for the 1-in-100 and 1-in-1,000-year events, respectively (Figure 6.13). Delays in flood peaks reached a maximum of five hours but varied with the nature of the event and antecedent conditions associated with rainfall and soil saturation.

In practice, it was not always possible to install permeable dams in locations identified by the initial model to be optimal because of practical constraints on the ground. These included lack of access for the construction equipment, steep, wooded or soft ground at the site prohibiting the use of heavy machinery, and farmers' concerns about waterlogging productive land or land used for vehicle access in winter. Although farmers accepted the concept of introducing permeable dams on their land, there was little sense of ownership and there were concerns about maintenance costs and liability in the event that failure of the dams caused damage downstream. On the other hand, at one site, a larger area of ground was made available to receive floodwater than had been assumed in the initial modelling process. This highlights the varying responses of individual farmers to proposals to install permeable dams, and the need to recognise factors other than simply production forgone, in itself something that is difficult to predict because of climate change uncertainty. Where payments to farmers are based on expected impacts on downstream flood risk and quantifiable damage to property, that uncertainty increases further.

Box 6.3 Leaky dams

Siting permeable dams carefully optimises both their performance and their longevity. Deep wide channels provide the greatest opportunity for water storage behind the dam, while natural basins in an adjacent floodplain provide additional potential for water storage outside the channel where this does not compromise land-use objectives. As well as minimising impacts on agricultural production from the land that is periodically flooded, the siting of out-of-channel storage should minimise impacts on vehicle access.

Large lengths of timber spanning the full width of the channel with an extension for keying into the bank on each side ensure that the dam is not eroded at either end. Avoiding bends in streams reduces the risk of bank scour associated with concentrated flow at the edge of the channel. In-channel posts should be avoided as these can cause debris to accumulate behind the dam, and if there is stream bed scour, they become destabilised, resulting in the collapse of the dam. The base of the dam should be well above the winter base flow level to optimise performance during major storm events and reduce the incidence of bed scour. Leaving a gap between lengths of timber also increases their permeability (Figures 6.14a, b and 16.5).

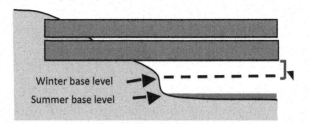

Figure 6.14a Location of permeable timber dam in relation to winter base flow

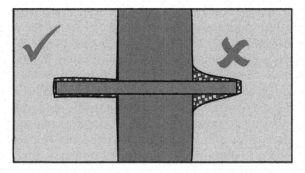

Figure 6.14b Keying timber into the bank to avoid erosion

Figure 6.15 Newly constructed permeable timber dam showing the use of larch tree
trunks and gap between the base of the dam and the winter base flow

Overview

The research described in this chapter has highlighted a number of important issues relating to the management of water. These have practical
implications for the management of catchments at a range of scales. It
has provided new information on the relative contribution of different
small headwater waterbodies to landscape scale species richness and the
transience of aquatic plant community structure. While there is clear potential for leaky dams to control downstream flood risk, our findings also
identify the limits of this approach when exposed to the most intense
storms, the frequency of which is expected to increase. Similarly, legacy
P, in both soils and sediments, will continue to have major impacts on
aquatic ecology which are likely to be exaggerated by low summer flow
conditions.

Our work has provided an insight into the relative contribution of domestic and agricultural sources of sediment and phosphorus and their impacts
on aquatic ecology at a range of scales. Domestic sources are linked to diet
and therefore to individual and societal dietary choice, one of the complex
topics discussed in Chapter 7. Agricultural contributions of sediment and
nutrients to water are also best addressed at source, as revealed by our
research into sediment traps and herbicide use. This requires catchment
scale collaboration or coordination of farmers with differing objectives,
worldviews, and social and economic circumstances, a subject explored in
Chapter 8.

References

Alsolmy, S. A. (2016) *The effects of agricultural land use on the community structure and functioning of small freshwater habitats.* Unpublished PhD thesis, University of Sheffield.

Arnold, J. G., Srinivasan, R., Muttiah, R. S. & Williams, J. R. (1998) Large area hydrologic modeling and assessment - Part 1: model development. *The Journal of the American Water Resources Association* 34, 73–89. DOI:10.1111/j.1752-1688.1998.tb05961.x

Berry, A. (2005) *Sediment sources in the river Eyebrook catchment using geochemical fingerprints and numerical mixing modelling.* Unpublished MSc thesis. Salford University.

Biggs, J., Stoate, C., Williams, P., Brown, C., Casey, A., Davies, S., Grijalvo Diego, I., Hawczak, A., Kizuka, T., McGoff, E., Szczur, J. & Villamizar, M. (2016) *Water Friendly Farming. Autumn 2016 Update.* Freshwater Habitats Trust and Game & Wildlife Conservation Trust, Oxford and Fordingbridge.

Brown, C. D., Turner, N. L., Hollis, J. M., Bellamy, P. H., Biggs, J., Williams, P. J., Arnold, D. J., Pepper, T. & Maund, S. J. (2006) Morphological and physiochemical properties of British aquatic habitats potentially exposed to pesticides. *Agriculture, Ecosystems and Environment* 113, 307–319.

Brown, C. D. & van Beinum, W. (2009) Pesticide transport via sub-surface drains in Europe. *Environmental Pollution* 157, 3314–3324. DOI:10.1016/j.envpol.2009.06.029.

Brown, C. & Velez, M. V. (2021) Unpublished data.

Crisp, D. T. & Carling, P. (1989) Observations on siting, dimensions and structure of salmonid redds. *Fish Biology* 34, 119–134.

De Bie, T., Stoks, R., Declerck, S., De Meester, L., Van De Meutter, F., Martens, K. & Brendonck, L. (2010) Agricultural Land Use Shapes Biodiversity Patterns in Ponds. In: Settele, J., Penev, L., Georgiev, T., Grabaum, R., Grobelnik, V., Hammen, V., Klotz, S., Kotarac, M. & Kühn, I. (Eds.) *Atlas of Biodiversity Risk.* 226–227. Pensoft, Sofia.

Defra (2008) *Linking agricultural land use and practices with a high risk of phosphorus loss to chemical and ecological impacts in rivers.* Research Project Final Report PE0116 and WT0705CSF. Department for Environment, Food and Rural Affairs, London.

Fleury, Z. & Perrin, C. S. (2004) Vegetation colonisation of temporary ponds newly dug in the marshes of the Grande Caricaie (lake of Neuchatel, Switzerland). *The Architectural Science* 57 (2–3), 105–112.

Foster, I. D. L. (2006) Lakes in the Sediment Delivery System. In. Owens, P. N. & Collins, A. J. (Eds.) *Soil Erosion and Sediment Redistribution in River Catchments.* CAB International, Wallingford. 128–142.

Foster, I., et al. (2008) *An Evaluation of the Significance of Land Management Changes on Sedimentation in the Eyebrook Reservoir since 1940.* Unpublished report.

Hamad, F. D. (2018) *The consequences of land management, particularly compaction, on soil ecosystems.* Unpublished PhD thesis. University of Leicester.

Hauer, C., Leitner, P., Unfer, G., Pulg, U., Habersack, H. & Graf, W. (2018) The role of sediment dynamics in the aquatic environment. In: Schmutz, S. & Sendzimir, J. (Eds.) *Riverine Ecosystem Management.* Springer, Cham. 151–169.

Jarvie, H. P., Withers, P. J. A., Bowes, M. J., Palmer-Felgate, E. J., Harper, D., Wasiak, K., Wasiak, P., Hodgkinson, R. A., Bates, A., Stoate, C., Neal, M.,

Wickham, H. D., Harman, S. A., & Armstrong, L. K. (2010). Streamwater phosphorus and nitrogen across a gradient in rural-agricultural land use intensity. *Agriculture, Ecosystems and Environment* 135, 238–252.

Lambert, S. J. & Davy, A. J. (2011) Water quality as a threat to aquatic plants: discriminating between the effects of nitrate, phosphate, boron and heavy metals on charophytes. *New Phytologist* 189 (4), 1051–1059. DOI:10.1111/j.1469-8137.2010.03543.x

Mitchell, C. (2010) *Sediment and nutrient dynamics at the Eye brook catchment, Leicestershire, UK*. Unpublished MSc thesis. Northampton University.

Morris, C. & Jarratt, S. (2016) *SIP – Sustainable Intensification Research Platform: Case Studies of Collaborative Initiatives*. University of Nottingham, Nottingham.

Neitsch, S., Arnold, J., Kiniry, J., Williams, J. & King, K. (2005) *Soil and Water Assessment Tool Theoretical Documentation Version 2005*. Texas A & M University, Texas.

Nicholls, R. (2015) *Quantifying fine-grained sediment storage in lowland headwater streams of the east midlands*. Unpublished MPhil thesis. University of Leicester.

Nishihiro, J., Nishihiro, M. A. & Washitani, I. (2006) Assessing the potential for recovery of lakeshore vegetation: species richness of sediment propagule banks. *Ecological Research* 21 (3), 436–445. DOI:10.1007/s11284-005-0133-y.

Ockenden, M. C., Deasy, C., Quinton, J. N., Surridge, B. & Stoate, C. (2014) Keeping agricultural soils out of rivers: evidence of sediment and nutrient accumulation within field wetland in the UK. *Journal of Environmental Management* 135, 54–62.

Owens, P. N. & Walling, D. E. (2002) The phosphorus content of fluvial sediment in rural and industrialised river basins. *Water Research* 36, 685–701.

Palmer-Felgate, E. J., Mortimer, R. J. G., Krom, M. D. & Jarvie, H. P. (2010) Impact of point-source pollution on phosphorus and nitrogen cycling in stream-bed sediments. *Environmental Science & Technology* 1, 44 (3), 908–914. DOI:10.1021/es902706r

RePhoKUs project (2021) Unpublished data.

Russell, M. A., Walling, D. E. & Hodgkinson, R. A. (2001) Suspended sediment sources in two lowland agricultural catchments in the UK. *Journal of Hydrology* 252, 1–24.

Stoate, C. & Sandford, D. (2008) *A survey of Brown Trout Salmo trutta fry in tributaries of the Eye Brook*. Game & Wildlife Conservation Trust, unpublished report.

Stoate, C., Brown, C., Vilamizar Velez, M., Jarratt, S., Morris, C., Biggs, J., Szczur, J. & Crotty, F. (2017) The use of a herbicide to investigate catchment management approaches to meeting Sustainable Intensification (SI) objectives. *Aspects of Applied Biology* 136, 115–120.

Villamizar, M., Stoate, C., Biggs, J., Morris, C., Szczur, J. & Brown, C. (2020). Comparison of technical and systems-based approaches to managing pesticide contamination in surface water catchments. *Journal of Environmental Management* 260. DOI:10.1016/j.jenvman.2019.110027

Villamizar et al. (in press).

Wasiak, P. H., Wasiak, K. A., Harper, D. M., Withers, P. J. A., Jarvie, H., Sutton, P. & Stoate, C. (2007) Impacts of agricultural land-use practices upon in-stream ecological structure and processes. *Plant Science* 130, 325–328.

Wasiak, P. (2009) *Studies on the effect of phosphorus upon headwater stream ecosystem structure*. Unpublished PhD thesis. University of Leicester.

Williams, P. J., Biggs, J., Barr, C. J., Cummins, C. P., Gillespie, M. K., Rich, T. C. G., Baker, A., Beesley, J., Corfield, A., Dobson, D., Culling, A. S., Fox, G.,

Howard, D. C., Luursema, K., Rich, M., Samson, D., Scott, W. A., White, R. & Whitfield, M. (1998) *Lowland Pond Survey 1996*. Department of the Environment, Transport and the Regions, London.

Williams, P., Whitfield, M., Biggs, J., Bray, S., Fox, G., Nicolet, P. & Sear, D. (2004) Comparative biodiversity of rivers, streams, ditches and ponds in an agricultural landscape in Southern England. *Biological Conservation* 115 (2), 329–341. DOI:10.1016/S0006-3207(03)00153-8

Williams, P., Biggs, J., Crowe, A., Murphy, J., Nicolet, P., Weatherby, A. & Dunbar, M. (2010) *Countryside Survey: Ponds Report from 2007. Technical Report No.7/07.* Pond Conservation and NERC and Centre for Ecology & Hydrology, Oxford.

Williams, P., Biggs, J., Stoate, C., Szczur, J., Brown, C. & Bonney, S. (2020) Nature based measures increase freshwater biodiversity in agricultural catchments. *Biological Conservation* 244 (108515) 1–14.

Withers, P. J. A. Hodkinson, R. A., Berberis, E., Presta, M., Hartikainen, H., Quinton, J., Miller, N., Sisak, I., Strauss, P. & Mentler, A. (2007). An environmental soil test to estimate the intrinsic risk of sediment and phosphorus mobilization from European soils. *Soil Use and Management* 23 (Supplement 1), 57–70.

Withers, P. J. A., Jarvie, H. P. & Stoate, C. (2011) Quantifying the impact of septic tank systems on eutrophication risk in rural headwaters. *Environment International* 37, 644–653.

7 Understanding and accepting complexity

Dichotomous belief is a binary thinking style that considers everything as being capable of division into two types, such as black and white or good and bad. This cognitive trait is increasingly prevalent in politics and religion (D'Antonio et al., 2013; Dutton, 2020). It is influenced by educational background but can also be associated with personality disorders (Mieda et al., 2020; Oshio, 2012). Social media accentuate polarised views on complex issues, sometimes contributing to conspiracy theory and dismissal of science (Bessi et al., 2015; Douglas et al., 2019). That is not to question the need for ideology. Ideology is important, even when driven by emotion, but it must be consistent with whatever scientific evidence is available if it is to contribute to our collective wellbeing.

There are numerous examples where the predominance of polarised thinking constrains understanding of agri-environmental issues and management (e.g.Ford et al., 2021; MacDonald et al., 2015). Understanding often complex issues is essential to meaningful discussion and planning effective action to address them. This is illustrated in this chapter by our intensive research into songbird ecology and conservation on farmland, and the principle can also be applied to many broader aspects of land management, as discussed later in this chapter.

Songbird conservation – habitat or predation pressure?

Nest survival

The predation of bird eggs or young from nests is a conspicuous activity that is clearly, at least potentially, detrimental to the productivity of individual birds. During the 1970s and 1980s, there was a sustained increase in the UK population of magpies *Pica pica*, one of the most active predators of bird eggs and young. Carrion crow *Corvus corone*, another avian nest predator also increased in numbers over the same period. The increases have been attributed to the feeding of birds in gardens, increased human waste, and the availability of game and wildlife carrion such as road kills associated with an increase in traffic over the same period (Pringle et al., 2019). Whatever the

DOI: 10.4324/9781003160137-7

reason, the combination of increased numbers of these conspicuous birds, and increased incidence of their conspicuous activity led to the assumption amongst many that they were the cause of declines in populations of songbirds that had occurred over the same period. The correlation seemed clear enough. This view was particularly strongly held by those involved in shooting who incorporated the control of crows and magpies into their game management systems.

On the other hand, many members of the wider conservation community drew attention to the habitat changes that had taken place over the same period, and in particular the intensification and simplification of farmland management. In terms of nest survival, the lack of management of hedges, an important nesting habitat for many species, was regarded as resulting in increased physical exposure to predators using visual cues (Dunn et al., 2016). In addition, the loss of insect food for breeding birds, and of seed food in winter, as discussed in Chapters 3 and 4, were considered to be major causes of population decline.

This potential role of nest predation on nest survival rates has been investigated at Loddington in relation to the legal control of nest predators as part of the game management system, and the collection of additional data from nearby sites where nest predators were not controlled. At Loddington, nest predators such as magpies, carrion crows, jackdaws *Corvus monedula*, foxes *Vulpes vulpes*, grey squirrels *Sciurus carolinensis* and brown rats *Rattus norvegicus* were controlled from 1993 (following a baseline year in 1992) to 2001, after which no removal of predators took place until 2011 when it was resumed at a lower level of intensity.

Throughout the early years of the Allerton Project, songbird nest monitoring was a major research activity, with 6,288 nests being monitored in the first decade. Most of the nests were of six species and percentage nest survival rates (mean ± se) differed considerably between them: blackbird *Turdus merula* (22.2 ± 2.18), song thrush *Turdus philomelos* (28.4 ± 3.74), dunnock *Prunella modularis* (51.1 ± 5.06), whitethroat *Sylvia communis* (63.6 ± 4.85), chaffinch *Fringilla coelebs* (33.1 ± 4.70) and yellowhammer *Emberiza citrinella* (40.9 ± 4.63) (Stoate & Szczur, 2001a). As well as having the highest nest survival rate, whitethroat also experienced the lowest percentage of nest failures attributed to predation (21.5%). Blackbird experienced the highest percentage of nests lost to predation (54.7%), followed by chaffinch (50.7%), song thrush (41.2%), yellowhammer (37.1%) and dunnock (30.2%).

There was a negative correlation between nest survival rate and carrion crow breeding density and for all species, this relationship being significant for blackbird, song thrush, dunnock and yellowhammer. A negative relationship between nest survival and magpie density was also apparent for all species except whitethroat and was significant for blackbird and song thrush. After allowing for habitat (nest height, exposure, etc.) and weather (rainfall and temperature) effects, the negative correlation between nest survival and corvid abundance was maintained for all species except for that between whitethroat nest survival and carrion crow abundance. Of the previously significant correlations, only that between song thrush and magpie ceased to be so. Small-scale use of imprint receptive dummy eggs in previously predated blackbird nests confirmed the prevalence of nest predation by corvids, with all but one of the fifteen nests being taken by these predators.

Although there appears to be a direct effect of crow or magpie abundance on nest survival rates of songbirds, for some species, the habitat in which nests are located has an equal or greater effect. Half the yellowhammer nests were located in field margin herbaceous vegetation, rather than in the hedge itself and nest predation rates were significantly higher for nests in hedges than those in herbaceous vegetation (Stoate & Szczur, 2001b). This is in line with the findings for whitethroat. This species had the highest nest survival rates and 95% of nests were located in herbaceous vegetation, rather than in hedges. The use of herbaceous field margin vegetation as a nest site is therefore associated with higher nesting success.

Within hedges, the vegetation structure and associated nest exposure may be influencing the probability of nest predation (Figure 7.1). We investigated the influence of habitat structure within hedges on nest survival rates of a range of hedge-nesting songbird species (Dunn et al., 2016). Birds selected nest sites with higher vegetation cover above the nest, increased visibility on the nest-side of the hedge, and reduced visibility on the far side. Nest survival rate at the nestling stage was higher where vegetation structure restricted access to corvid-sized predators, and at nests close to potential vantage points. Nest survival during the nestling stage was significantly higher in stock-proof, mechanically trimmed hedges than in recently laid

Figure 7.1 Song thrush nest in holly bush

or neglected open-structured hedges. In trimmed hedges, nest survival increased with time since the last trim over a four-year period. So hedges that are cut regularly, but not every year or every other year, are associated with higher nest survival because of the better concealment they afford nests.

There is also a seasonal effect on nest exposure. When crows and magpies were abundant, early-season blackbird nests were more susceptible to predation than nests later in the season when they were better hidden by leaves (White et al., 2008). Removal of crows and magpies as part of the game management at Loddington brought early-season nest survival rates up to those of mid-season nests and resulted in higher overall annual productivity in terms of young birds present towards the end of the nesting season (White et al., 2008).

Breeding bird abundance

Spotted flycatcher *Muscicapa striata*, a species that declined by 88% in the period 1970–2018 and that is susceptible to nest predation, also experienced a habitat effect on nest survival rates, this time at the breeding territory scale (Stoate & Szczur, 2006). For this species, reducing the number of nest predators had a significant effect on nest survival rates to fledging: 77% when predators were removed and 16% when they were not. However, in the years when predators were abundant, nest height had a positive effect on survival rates, and nest predation was also significantly higher in woodland than in village gardens. There was also a response in terms of breeding abundance. Spotted flycatcher breeding numbers doubled in the 1993–2001 period when predators were removed, and then reverted rapidly to the original numbers in 2003 in the absence of predator control. Changes in territory numbers took place mainly in woodland where nest predation was highest (Figure 7.2). For this species, at least, there appears to be an effect of both habitat and predator abundance on both nesting success and breeding numbers in subsequent years.

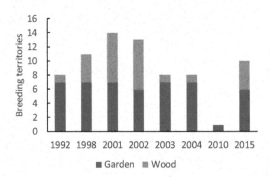

Figure 7.2 Number of spotted flycatcher territories in woodland and gardens at Loddington. Predator control was carried out from 1993 to 2001, and again from 2011

This was also the case for blackbirds studied by White et al. (2008); as well as increases in nest survival during the period in which crows and magpies were removed, there was also an increase in blackbird breeding numbers, and this declined when removal stopped, and corvid numbers increased again. Of the other five species whose nest survival was monitored, song thrush, dunnock, whitethroat and chaffinch all increased in abundance between 1993 and 1998, only yellowhammer failing to do so (Stoate & Szczur, 2001a). A comparison of the abundance of these species at Loddington and in the surrounding area was carried out over a three-year period, starting in 1995, the third successive year in which predators were controlled (Stoate & Szczur, 2001a). Breeding densities were generally higher at Loddington than in the surrounding area, with the greatest differences being reported for blackbird and song thrush. For whitethroat, the difference was only apparent in 1997. Overall, considering all songbird species present at Loddington, those species that had been declining nationally over the previous 25 years increased in abundance at Loddington more than species that had maintained relatively stable populations nationally.

We compared changes in bird numbers at Loddington over a 19-year period from 1992 with equivalent changes at a farm managed by the Royal Society for the Protection of Birds (RSPB) in Cambridgeshire over an 11-year period (Aebischer et al., 2016). The Cambridgeshire farm was primarily open farmland, whereas Loddington comprises a mix of arable and grass farmland, hedges and woodland. The bird community reflects this difference, with farmland specialists being relatively more numerous at the RPSB farm, and open-cup nesters, especially passerine insectivore being more abundant at Loddington. Management strategies at both farms were successful in increasing breeding numbers of birds. The maximum annual rate of increase was nearly 20%, and for farmland species, it was 12% at both sites. For Biodiversity Action Plan species that had declined nationally over previous years, the annual rate of increase was 50% higher at Loddington than at the RSPB farm, and it was twice as high for non-passerine herbivores, passerine granivores and open-cup nesters. In all cases, the increases in bird numbers substantially exceeded underlying regional trends which were more or less stable over the same period.

This exercise was particularly informative in highlighting the influence of landscape character on susceptibility of nests to predation as the RSPB farm comprised a more open landscape with fewer hedges and little woodland. Heterogeneous landscapes such as that at Loddington support more songbird species than homogeneous open farmland landscapes; 72 bird species were recorded at Loddington for example, compared to 55 at the RSPB farm. But heterogeneous landscapes also support more nest predators such as carrion crows and magpies. When their numbers were not controlled, these corvids were up to four times as abundant at Loddington as they were at the RSPB farm. The removal of nest predators therefore increases nest survival rates of vulnerable species in complex landscapes but not in simple ones where numbers are already inherently low. Where predator numbers are low, habitat management alone can potentially result in increases in breeding numbers of species associated with these open homogeneous landscapes.

In the early years of the Allerton Project (1993–2001), predator control has also influenced numbers of pheasants *Phasianus colchicus* (Figure 7.3), grey partridges *Perdix perdix* and brown hares *Lepus europeaus* (Figure 7.4). The employment of a gamekeeper over this period was intended to provide a demonstration of best game management practice, explore the potential of the game management system for non-game bird conservation, and provide an autumn surplus of wild-bred game birds for shooting. This was achieved, with eight to ten small shoots being held each winter. The size of the hare and wild pheasant populations declined dramatically from 2001 when the predator control element of the management system was withdrawn in order to document responses in both game and non-game birds (Figures 7.3 and 7.4). Overall songbird numbers also declined but to a lesser extent. From 2007, the provision of food in winter was also stopped, and overall songbird numbers declined further, while gamebird numbers remained at a low level. From 2011, a small shoot was again established, but this time based on reared and released pheasants, with a lower level of predator removal.

During the period of wild game management and subsequent cessation of predator control and winter feeding, autumn numbers of pheasants were correlated with abundance of breeding songbirds the following spring, suggesting a benefit of these game management activities to songbird conservation (Figure 7.5). However, this relationship broke down in the subsequent

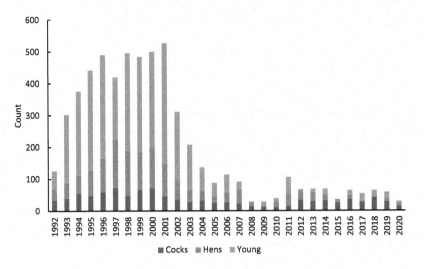

Figure 7.3 Numbers of male (cocks), female (hens) and young pheasants at Loddington in autumn. Predators were controlled from 1993 to 2001, and again from 2011

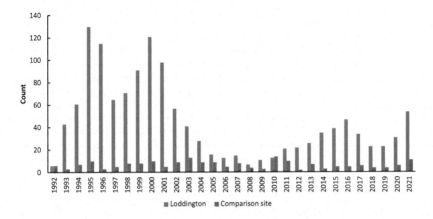

Figure 7.4 Winter brown hare numbers at Loddington and at a comparison site (Hallaton) with no game management. Predators were controlled at Loddington from 1993 to 2001, and again from 2011

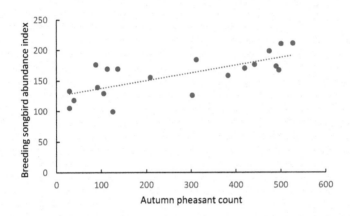

Figure 7.5 Relationship between autumn pheasant numbers and breeding songbird numbers (1992–2011)

period in which habitat management and winter feeding were carried out, but with reduced emphasis on control of nest predators as, while songbird numbers increased, the nesting success and autumn numbers of pheasants were low.

Observations of species exploiting feed hoppers as a source of winter food revealed that these species included nest predators such as corvids, brown rats and grey squirrels. While the provision of supplementary food in winter may be improving winter survival of some songbird species, this may also be the case for some potential predators. It has also been suggested that the releasing of reared gamebirds may be contributing to the survival or productivity of nest predators through the increase in prey and carrion available to them (Pringle et al., 2019).

In 2018 and 2019, 65 hen pheasants were radio-tracked from March to early July in order to understand their survival and nesting success. The 65 hens produced only nine nests, all of which failed, the majority being predated. All but five of the hen pheasants died during the study period. Seventy-eight percent of these were predated or scavenged. It was difficult to distinguish between the two outcomes as some of the birds were known to have died or been weakened by diseases such as coronavirus, Heterakis, gapeworm and egg peritonitis (Sage, 2020). As well as causing direct mortality and subsequent scavenging by predators, this could also make pheasants less able to breed and more susceptible to predation.

A simple examination of changes in abundance of various songbird species over the 30 years of monitoring illustrates how different processes impact differently on these species. It is important to understand that these processes may operate at the landscape, national or even international scale, influencing the extent to which management on an individual farm can affect bird abundance. For blackbird and song thrush, two ecologically similar species whose nest survival we have demonstrated to be affected by predator abundance, breeding numbers increased during the 1993–2001 period of wild gamebird management including the control of carrion crows and magpies. They then declined, along with nest survival rates, when these predators were not controlled (2002–2010), and increased when the full suite of game management was resumed (Figure 7.6). Detailed territory mapping reveals that in 2021, song thrush and blackbird numbers were, respectively, 44 and 121, compared to 14 and 66 in 1992.

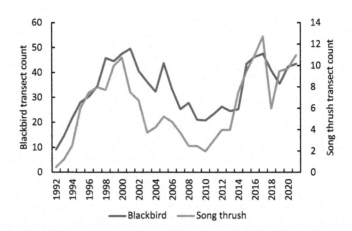

Figure 7.6 Blackbird and song thrush breeding abundance at Loddington based on annual transect data

Other species showed a similar trend to that revealed for these two thrushes, with linnet (*Carduelis cannabina*) and bullfinch (*Pyrrhula pyrrhula*) providing two examples of finches. Detailed mapping of breeding territories every five to six years shows that the number of linnet territories in 2021 was 23 compared with 10 in 1992, and bullfinch numbers had increased from 6 to 20. However, not all finches followed the same pattern. Two other ecologically similar species, chaffinch and greenfinch *Carduelis chloris* showed a very different trend to the two thrushes (Figure 7.7). Chaffinch was one of the species whose nest survival rates were affected by predator abundance, and this species also exploited supplementary food provision in winter. However, our monitoring reveals that, despite an initial increase in numbers, and a slight decline when predator control ceased, there was then a more sustained decline in abundance from 2006. Although much less numerous, greenfinch abundance reflected the same trend. Both species are susceptible to Trichomonosis, a protozoan parasite which is thought to have been transferred from pigeons to finches, causing substantial chaffinch and greenfinch mortality and population declines across the UK from 2005 (Robinson et al., 2010). In this case, impacts acting on these species at the national level prevented any response to management at the farm scale. In fact, territory mapping reveals that, from being the most numerous breeding passerine with 135 territories in 1992, chaffinch numbers had declined to 72 by 2021.

Yellowhammer is an iconic farmland bird species which has suffered a sustained and continuing population decline nationally (Defra, 2019). As well as the impacts of habitat and predator abundance referred to above, this is one of the species to make use of wild bird seed crops and supplementary feeding through the winter, and the number of breeding territories in 2021 was 55, only marginally below the 57 recorded in the baseline year of 1992.

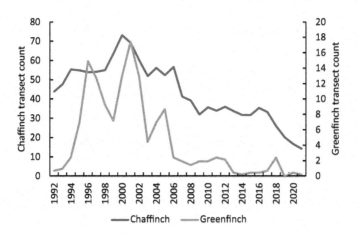

Figure 7.7 Chaffinch and greenfinch breeding abundance at Loddington based on annual transect data

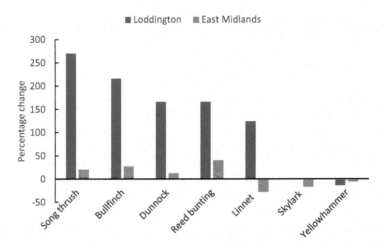

Figure 7.8 Change in breeding abundance of nationally declining songbird species for which comparable data are available for Loddington (1992–2021 territory mapping data) and the East Midlands (1995–2018 Breeding Bird Survey data) (BTO, 2019)

Of the species that have declined nationally over the past 30 years, those that have shown a subsequent slight increase in abundance in the East Midlands increased very substantially at Loddington in response to the management (Figure 7.8). Skylark and yellowhammer, which showed slight declines in the region, did not increase in abundance at Loddington. Loddington may be acting as a source population for these two species with young skylarks and yellowhammers dispersing to occupy vacated territories on adjacent farms, while the young of species such as song thrush, bullfinch and reed bunting remain on the farm, contributing to an increase in breeding abundance (Table 7.1).

Winter conditions for migratory birds

Of the warblers, whitethroat was the most intensively studied, and as stated previously, exhibited the highest nest survival rates, even in the presence of high predator numbers. It is well established that whitethroats are one of a number of species whose breeding numbers in northern Europe are determined largely by conditions on the sub-Saharan wintering region and migration route (Johnston et al., 2016). Breeding numbers in the UK reflect conditions such as drought during the winter and northerly winds and dust storms in the spring. The northwards migration in the spring requires accumulation of fat towards the end of the dry season on the southern edge of the Sahara, prior to crossing to northern Africa when wind conditions are right.

Table 7.1 Management changes and numbers of songbird breeding territories recorded at Loddington for Biodiversity Action Plan species

	1992	1998	2001	2006	2010	2015	2021
Habitat management		X	X	X	X	X	X
Predator control		X	X			X	X
Winter feeding		X	X	X		X	X
Skylark	36	36	37	33	26	37	35
Yellow wagtail	3	5	3	1	2	6	1
Dunnock	46	86	144	97	51	135	111
Song thrush	14	48	64	34	15	44	60
Spotted flycatcher	8	11	14	6	1	10	6
Marsh tit	4	4	7	3	No data	8	6
Willow tit	1	1	1	1	1	2	0
Tree sparrow	3	0	7	18	12	7	2
Linnet	10	21	25	17	15	22	23
Bullfinch	6	11	12	6	12	18	20
Yellowhammer	57	55	54	46	41	44	55
Reed bunting	3	3	3	5	8	5	13

Laying down fat is achieved by increasing fruit consumption. Stoate and Moreby (1995) demonstrated how the berries of *Salvadora persica* (Figure 7.9) contributed to pre-migratory fattening of whitethroats and other *Sylvia* warblers in the inundation zone of the River Senegal in 1993, and that larger berries (>6 mm diam.) were selected by the birds. Very substantial defoliation of *Salvadora* bushes by desert locusts *Schistocerca gregaria* in 1994 enabled a comparison to be made with the previous year. Inflorescences had partially regrown by the time of the whitethroat migration but their abundance was less than 20% of that in the previous year, and berry size was significantly smaller (Stoate, 1995). Whereas 49% of the whitethroats present in 1993 had high fat scores, in 1994, only 14% did so, reducing the likelihood of successful crossing of the Sahara on the northward migration. Although, in this case, reduced berry abundance was the result of defoliation by locusts, drought, excessive browsing by cattle, or clearance of bushes for agricultural purposes are all additional pressures that reduce habitat and food availability to migratory birds.

Elsewhere in West Africa, livestock browsing and clearance of trees and shrubs for cropping are also threats to birds, including migratory species that over-winter in the region. *Acacia* species such as *A. albida* and *A. nilotica* support potential invertebrate food for warblers and are particularly important tree species for migratory birds (Jones et al., 1996; Stoate, 1997). *A, albida* is unusual in that it is leafless for part of the year, coming into leaf in December, but supports an abundance of invertebrate food for birds during the dry season.

A comparison of the use of grazed and relatively ungrazed *Acacia* woodland (Figure 7.10) by wintering olivaceous warblers *Hippolais pallida*, in

Figure 7.9 Berries of *Salvadora persica* in the floodplain of the Senegal river, northern Senegal

Figure 7.10 Acacia woodland close to the northwestern Gambia/Senegal border

November and February highlights the ecological implications of livestock in savanna woodland (Stoate, 1998). Sixty-five percent of trees in the ungrazed area supported climbing plants and olivaceous warbler density in November was significantly correlated with the canopy area of *Acacia* trees with climbers, but not with total canopy area. This is explained by the fact that potential invertebrate food for the warblers was significantly more abundant in trees with climbers than in those without. The positive effect of climbing plants on invertebrates and birds was not apparent in February when the *Acacia* trees were in leaf, and although invertebrate numbers were higher in February, there was no longer a difference between trees with and without climbers. Relatively undisturbed woodland that supports climbing plants therefore increases the availability of food in the early part of the wintering period, potentially influencing winter survival.

Where trees have been cleared for crop production, few trees remain and the woody vegetation is represented mainly by shrubs. On agricultural land on the northern Gambia/Senegal border, although some trees were present at low densities, no whitethroats were recorded using them, but whitethroat density was correlated with overall shrub cover, and with cover of the most common species, *Guiera senegalensis* at 1.02 ha^{-1} (Stoate et al., 2001). *G. senegalensis* was also the shrub species supporting the highest numbers of potential bird food invertebrates, most notably caterpillars and spiders. The whitethroat density present in this study area was similar to that of 1.1 ha^{-1} in northern Nigeria where *Piliostigma* shrub species were associated with high invertebrate abundance (Stoate, 1997). Even where trees have been cleared, shrubs such as these perform an important ecological role in supporting some of the migratory birds that breed in northern Europe. The management of those shrubs and trees will depend largely on their perceived value to farmers for agricultural uses as well as medicine, fuel, timber and cultural uses or values (Stoate et al., 2001).

Two ecologically similar migratory warblers that shed further light on the complexity of these international impacts are chiffchaff (*Phylloscopus collybita*) and willow warbler (*P. trochilus*). Despite occupying similar ecological niches, their population trends at Loddington have been very different (Figure 7.11). Although the processes involved are not fully understood, the difference may be because of the different wintering ranges occupied by the two species. Chiffchaff which has increased in abundance 15-fold over the 30 years of monitoring has a more northerly wintering range, with increasing numbers of birds remaining in the UK as winters have become milder over this period. Willow warblers winter further south in sub-Saharan Africa. Their numbers at Loddington have declined substantially, in line with the same population trend for the rest of southern England. However, willow warbler numbers in Scotland have been increasing over the same period, along with those of chiffchaff. This demonstrates that the changes in abundance further south are not a result of competition between the two species. Willow warblers breeding in Scotland are likely to

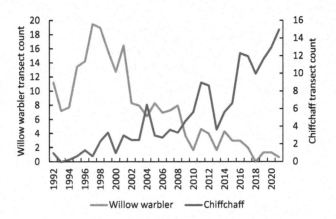

Figure 7.11 Chiffchaff and willow warbler breeding abundance at Loddington based on annual transect data

be wintering in different areas to those from further south in the UK, with differing conditions in those wintering areas influencing winter survival or fitness (Morrison et al., 2013). There is a general trend for species such as whitethroats which winter in the Sahelian region of sub-Saharan Africa to have experienced population declines earlier than species such as willow warbler wintering in the humid tropics (Vickery et al., 2013).

The UK is also a wintering area for many bird species which breed elsewhere in Europe. A classic example is woodcock *Scolopax rusticola* which is associated mainly with the larger woods around Loddington. Very low-level shooting of woodcock was carried out in one of the woods, with the rest of the woodland being left as a sanctuary with no shooting of woodcock. The scale of the shooting was dependent on numbers of woodcock present, with standardised counts being conducted in late November and early December. This low level of shooting has enabled us to learn a considerable amount about the breeding range of the woodcock wintering at Loddington as feathers from each of the shot birds were collected for stable hydrogen isotope analysis. This identifies the geographical area in which the birds were reared. Woodcock wintering at Loddington come from an area extending from Germany, through the Baltic States and Poland to Russia (Figure 7.12; Andrew Hoodless, unpublished data, 2021). The evidence available from eastern Europe suggests that woodcock breeding success and abundance are stable in this region.

The issue of whether predator numbers or habitat type most influence songbird nesting success and abundance is clearly not a binary issue, with multiple interacting factors affecting different species in different ways. But the research that has been carried out since 1992 helps us to understand this and to make more informed choices about future management.

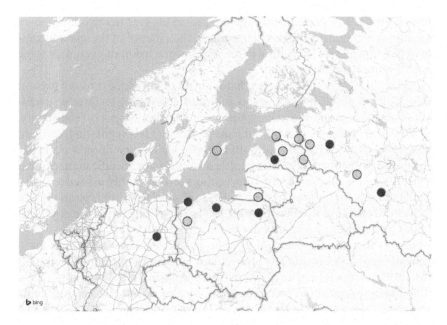

Figure 7.12 Estimated breeding range of woodcock wintering at Loddington based on stable hydrogen isotope analysis of feathers (Hoodless, 2020)

The examples provided for bird conservation illustrate the need to understand complexity and interaction at a range of scales; this principle applies to many other aspects of agri-environmental land management.

Farming or rewilding

Chapters 3–5 have been about farmland management and a range of interventions intended to improve food production, the sustainability of that production, and in particular, the conservation of wildlife associated with productive agricultural land. Diametrically opposed to this approach to land management is rewilding. Originating in America, rewilding was, and continues to be, characterised by large core protected areas, the ecological connectivity between them, and the role of keystone species, especially large predators.

This approach, comprising very large, interconnected areas, and free movement of large herbivores and carnivores did not translate well to Europe. The Oostvaardersplassen project in the Netherlands required fencing to constrain the movement of large herbivores within the 5,600 ha area. The presence of the fence and the lack of large carnivores resulted in a rapid increase in herbivore numbers, large-scale destruction of regenerating vegetation, and starvation of large numbers of animals (Kopnina et al., 2019). Culling was subsequently introduced to control herbivore numbers and allow regeneration of the vegetation. In the absence of the space available in North America, rewilding in most of Europe inevitably requires some form of human intervention.

Rewilding is a dynamic process that requires disturbance through natural biological and physical processes to achieve a continuing long-term succession of vegetation and the habitat heterogeneity that supports biodiversity. Such natural processes provide a means of creating self-regulating, resilient ecosystems (Pettorelli et al., 2018). In Europe, such intervention is normally introduced in the form of grazing cattle and sheep, or specific management by people and machines. These interventions are normally associated with specific targets for habitat, or even individual species, in contrast to the objectives of rewilding to allow natural processes to determine the course of habitat and species community development.

There is therefore a spectrum of approaches to conservation land management involving varying levels of human intervention, ranging from various forms of rewilding, or perhaps more appropriately, 'wilding' (Tree, 2018), to management interventions on productive farmland to meet specific biodiversity objectives. In the words of Fuller and Gilroy (2021), "There is nothing inherently problematic about valuing beautiful and species-rich elements of the countryside that are created and maintained by the interventions of people".

Especially in the lowlands of the UK and continental Europe, management that is considered as contributing to rewilding involves grazing and browsing by normally primitive domesticated livestock whose numbers are controlled and normally harvested for meat, as well as other human interventions such as fencing and direct management of vegetation and water levels.

The approach is also carried out at a range of scales. On many farms, small areas of land are left for natural regeneration of trees. At Loddington, a 3 ha area of former riparian arable land that is susceptible to flooding was entered into the first voluntary set-aside scheme in 1991 and has received only limited management since then, except for the creation of a network of small ponds to intercept arable runoff and create wildlife habitat (Stoate et al., 2007; Figure 7.13). The diversity of habitats, from small ponds and waterlogged areas, through open grasslands to naturally regenerated stands of alder *Alnus glutinosa*, supports a wide range of wildlife species, including those occurring on other parts of the farm and some that are unique to the site. The naturally regenerating site therefore contributes to the enhancement of landscape scale biodiversity.

The pools supported 85 aquatic macro-invertebrate species, ranging from 24 to 52 species per pool and including six nationally scarce Coleoptera.[1] The pools also supported a total of 30 macrophyte species, ranging between 9 and 18 per pool. Fifteen species of Odonata have been recorded. Terrestrial

Figure 7.13 Low lying area of riparian land at Loddington with naturally regenerating vegetation

invertebrates included five Orthoptera species that were previously absent
from the site, including three species that were not recorded elsewhere on
the farm. Common and slender groundhoppers *Tetrix undulata* and *T. sub-
ulata* were abundant on the drying mud of pond margins. Lesser marsh
grasshopper *Chorthippus albomarginatus*, Roesel's bush-cricket *Metrioptera
roeselii* and long-winged conehead *Conocephalus dorsalis* occurred in taller
vegetation. Twelve species of hoverfly (Diptera, Syrphidae) that have a close
association with aquatic habitats were recorded (John Szczur, pers, comm.).
Three of these, *Chrysogaster cemitiorum*, *Helophilus hybridus* and *Parhelo-
philus versicolor*, are locally scarce. Large marsh horsefly *Tabanus autumna-
lis* was recorded on marginal pond vegetation and is a species that is scarce
away from the coast in England and had not previously been recorded in
Leicestershire.

This provides an example of how taking land out of production and combining only limited habitat creation and management with natural processes such as flooding and regeneration of vegetation can quickly benefit
biodiversity. Views differ amongst academics on the relative merits of taking
land completely out of production for wildlife conservation like this while
using other land more intensively (land sparing) and meeting conservation
and food production objectives simultaneously on the same area of land

through integration (land sharing). While such 'land-sparing' segregation of productive land and areas for wildlife conservation at the scale of the North American rewilding areas is not feasible at the same level in lowland England, the riparian land at Loddington provides an example of how such an approach can be beneficial at a small scale within farms.

Land sparing at an even smaller scale, in field margins, as described in Chapter 5, can result in benefits to food production through natural control of crop pests and pollination of crops and is therefore arguably more akin to land sharing. It is questionable to what extent the terms that are currently in use to create a dichotomous approach to land management are helpful in lowland European landscapes where such integration tends to predominate.

Arguably, rewilding that combines low-density grazing and browsing by domestic ruminants which are harvested for meat is more characteristic of the integration of agricultural production and wildlife conservation through 'land sharing', albeit towards the land-sparing end of the spectrum. In practice therefore, it is rarely possible, or desirable, to adopt exclusively land sharing or sparing approaches. The need to integrate both approaches at a range of scales, and incorporate networks to benefit species that cannot exploit agricultural land, as well as those that can, is increasingly accepted (e.g. Grass et al., 2019).

Silvopasture, which combines trees and livestock at a range of intensities, provides an example of low-level land sharing that benefits biodiversity alongside food production. While forestry and agriculture have long been distinct in both policy and practice, there is an increasing recognition that there is scope for breaking down this dichotomy and encouraging their integration.

Farming or forestry

The Oostvaardersplassen rewilding project example illustrates the conflict between tree cover and ruminant density as the natural regeneration of trees is halted above a certain threshold of livestock density. In the uplands of Scotland, similar constraints on the natural regeneration of trees are created by high densities of red deer (Carver & Convery, 2021), but low livestock density silvopasture systems are widespread across much of Europe with trees providing benefits to livestock in the form of shelter, shade and supplementary forage. In the UK, benefits of trees to livestock welfare and liveweight gain have been demonstrated for the Welsh uplands (He et al., 2017). Such findings challenge the strongly established polarisation between 'forestry' and 'farmland' which has been accentuated by division of these two land uses within long-term government policy, and by economic pressures to specialise production.

Historically, some tree species were pollarded to provide fodder during periods of food shortage. Tree leaves can also provide a supplementary source of minerals where the availability of these is low in some fields, or during periods in the grazing season. Cobalt is a particularly important mineral for ruminants as it is associated with the synthesis of vitamin B_{12} and deficiencies can severely limit growth. Research at Loddington has explored spatial differences and seasonal changes in this and other minerals in grazed sward (Lee et al., 2017).

Simulated grazing samples were collected on 17 occasions from 15 fields across the Allerton Project farm and a neighbouring farm across two grazing seasons (2015, 2016) and these were analysed for trace element composition. Blood samples were taken from sheep to compare their trace element status with the concentrations in the grazed pasture. The selenium status of all fields was generally low with many not even meeting the lower ewe maintenance requirements, especially in May, June and July across both years and for August and September in 2016. Zinc and manganese concentrations in the sward were correlated, with some but not all of the permanent pasture fields being higher for both, while the short-term leys and clover fields were generally lower. Manganese was above sheep dietary requirements for all fields, while zinc concentrations were between the upper and lower range of requirements and therefore some sheep were likely not to meet zinc requirements from grazing alone.

Cobalt concentrations in the sward varied between fields and seasonally, and were above the 0.2 mg/kg of dry matter required by growing lambs only at each end of the grazing period. Rainfall affected concentrations so that sufficient cobalt was available in April 2016 when there was higher than average rainfall (~110% of 1981–2010 rainfall average), but only reached 0.1 mg/kg in the dry spring of 2015 (~70–90% of 1981–2010 rainfall average). In both years, the concentration was below 0.1 mg/kg through the summer period from June to September, and cobalt concentrations in blood samples from grazing sheep reflected the availability in the sward during this period. Cobalt in grazed sward was therefore below the requirements of growing lambs and some ewes for much of the summer and there was a subsequent ten-fold increase in cobalt concentration in the sward between August and November (Figure 7.14).

Pastures containing white and red clover had higher cobalt concentrations than predominantly ryegrass swards. Seasonal reductions in the availability of cobalt can therefore be addressed by including clover in the sward, while an understanding of spatial variability in cobalt concentrations across fields can guide whether grass is grazed or harvested as forage and subsequently fed to housed animals with mineral supplements. However, the availability of tree fodder that is high in cobalt or other minerals could represent another means of addressing both seasonal and spatial shortfalls in sward minerals.

To investigate this, along with other research partners, we collected leaf samples from goat willow *Salix caprea*, common alder and pedunculate oak *Quercus robur* trees at Loddington, at Bangor University's Henfaes research farm, and at the Organic Research Centre's farm in Berkshire and analysed the leaves

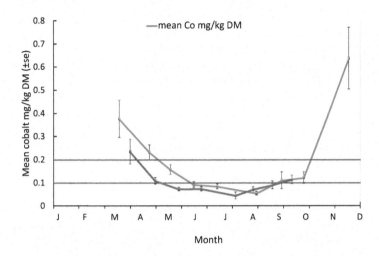

Figure 7.14 Mean cobalt concentrations (mg/kg DM) for grazed sward in 2015 (red) and 2016 (blue). The horizontal lines indicate the recommended cobalt requirements for sheep maintenance (bottom) and growth (top) (Kendall, 2017)

for a range of minerals (Kendall et al., 2021). Zinc concentrations were higher in willow leaves than in alder and oak, while alder had higher concentrations than oak. Cobalt concentrations were higher in willow than the other two species, and the concentration of selenium was affected more by the site than the tree species. Other trace elements were found to occur within the requirement range for sheep, potentially offering similar concentrations as sward.

There were also differences in metabolisable energy and crude protein. Alder had higher concentrations of metabolisable energy than oak, while oak had higher concentrations than willow. Crude protein content was higher in alder than in oak although willow was not found to differ from either of the other species. There was a general tendency for crude protein to be higher in leaves sampled in June than in those sampled in September.

There are therefore potential nutritional benefits of feeding tree leaves to ruminants, such as a supplementary source of zinc. In particular, the high concentrations of cobalt in willow leaves in the summer coincides with the period in which cobalt concentrations in grazed sward are normally declining below the requirements of grazing animals, especially growing lambs. We therefore designed a simple experiment to test whether feeding willow leaves to weaned lambs would result in elevated blood cobalt and vitamin B_{12}.

Two groups of six lambs were fed 300 g of willow leaves per day over a two-week period. Over the same period, two groups of lambs were not fed willow. Both groups had ad lib access to pasture and water. Blood samples were taken at the start and end of the two-week period. Blood cobalt concentrations doubled and vitamin B_{12} was 2.6 times higher in willow-fed lambs than in those that were not fed willow leaves (Walker, 2019). In the following year, a third group was fed willow at the start of the two-week period and again after one week, in addition to a daily-fed group and a control group that was not fed willow. Significantly elevated blood cobalt concentrations were again found in the daily-fed group but the cobalt concentration in the weekly fed group was only marginally higher than the control group. This implies that continuous access to willow is needed in order to realise the benefits.

Willow-fed livestock may also reduce the impact of red meat production on climate change. Nitrous oxide emissions from ruminant urine patches are an important component of the agricultural contribution to greenhouse gas emissions which amounts to around 70% of total UK emissions of this gas. Most of the agricultural contribution comes from soils as a result of mineral or organic fertiliser application and deposition of dung and urine by grazing animals. Urine contributes approximately 60% of the nitrogen excreted by grazing animals (Chadwick et al., 2018). Condensed tannins that are present in tree leaves have the potential to suppress microbial activity in the rumen, reducing the uptake of nitrogen into the blood, and ultimately into urine. This has the potential to reduce emissions of nitrogenous gases, primarily nitrous oxide and ammonia from urine patches. Inhibition of microbial activity in the soil could have the same effect. As nitrous oxide is a major greenhouse gas, and ammonia has negative air quality implications, the use of willow to reduce these gaseous emissions from urine could potentially contribute to both climate change and human health targets.

In August 2020, we fed an average of 200 g of goat willow leaves per day to two groups of six weaned lambs over a two-week period (Stoate et al., 2021). Another two groups of six lambs were not fed willow. At the end of this period, we identified fresh urine patches by direct observation of the lambs (six willow-fed and six not willow-fed) and used our Gasmet gas analyser to measure emissions of carbon dioxide, as well as nitrous oxide and ammonia. We did this within 20 minutes of urination, and again one and two weeks later.

There was a consistent trend for urine patches in pens with lambs that were fed willow having lower emissions than those that were not fed willow for each of the three gases, although this was only statistically significant for nitrous oxide in Week 2 as the sample size was low. Ammonia emissions declined rapidly, nitrous oxide emissions were mainly in Week 2, and carbon dioxide emissions declined gradually over the two-week period. Lower carbon dioxide emission suggests that microbial activity was suppressed in the soil, rather than in the rumen, but we cannot discount a contribution from the latter.

We repeated this experiment in 2021, collecting gaseous emission data daily for two weeks from nine urine patches in the groups that were fed willow daily and those that were not fed willow, and additional data from pasture where there was no urine patch as a control. We obtained similar results

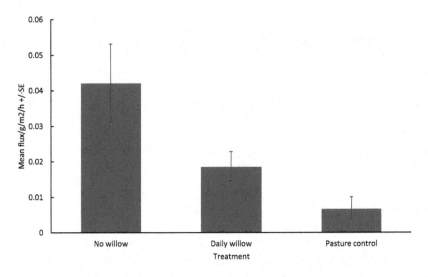

Figure 7.15 Three-day mean nitrous oxide emissions from urine patches (lambs fed willow daily, and not fed willow) and from pasture without urine (control)

to the previous year, except that there was no difference in carbon dioxide, suggesting suppression of microbial activity in the rumen rather than in the soil. Nitrous oxide emissions were significantly lower for the group that was fed willow for the first three days (average 56% reduction), after which there were low emissions from both groups (Figure 7.15). Giving lambs access to willow could therefore make a valuable contribution towards reducing the global warming potential of red meat production.

Condensed tannins in willow leaves have also been shown to inhibit the development of larval intestinal worms in ruminants. This has important ecological implications as anthelmintics can have substantial negative impacts on dung beetles which have an intrinsic conservation value, provide food for other wildlife, and perform an important ecological function in terms of the breakdown of dung and recycling of nutrients. In 2021, a lab-based assay examined the potential of tannins extracted from goat willow leaves to control larvae of the parasitic nematode, *Cooperia oncophora* (Elsheikha, 2021; Figure 7.16). The larvae were exposed to the tannin at a range of concentrations for 24 hours at 27°C. After this period, nearly half of the larvae that were exposed to the 0.5% solution were dead, while at 2% concentration, 95% had died (Figure 7.17).

It is already increasingly recognised that trees provide important shade and shelter for livestock (He et al., 2017). Our results suggest that ruminant access to willow leaves could supplement minerals, reduce greenhouse gas and ammonia emissions from urine patches, and reduce intestinal worm

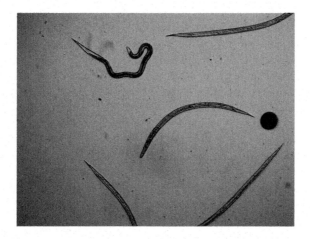

Figure 7.16 Larval *Cooperia oncophora* nematodes used in our bioassay. Photo courtesy of Hany Elsheikha, University of Nottingham

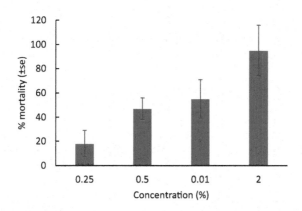

Figure 7.17 Mortality of nematodes in relation to extracted tannin concentration (Elsheikha, 2021)

burdens and the use of anthelmintics which reduce grassland invertebrate populations.

Together, these findings make a strong case for the integration of trees into ruminant livestock systems, in contrast to the prevalent practice of isolating livestock from trees. There is also limited but growing interest in the potential role of trees in arable systems, although historically, tree cover is more strongly segregated from arable cropping than it is from livestock. The benefits are also additional to those associated with carbon sequestration and catchment scale management of water quality and flood risk (Chapter 6), biodiversity, food and timber.

Other false binaries within agricultural systems

There are now well-established distinctions between organic and what have become known as 'conventional' farming, although 'conventional' farming systems have only been conventional over the past 70 years or so. The dichotomy is accentuated by the need for organic farming to be certified, requiring strict adherence to the required criteria to qualify for organic status and achieve price premiums.

Environmental benefits, including enhanced biodiversity often, but not always, result from organic farming (Lichtenberg et al., 2017), but 'conventional' farming systems have the flexibility to adopt a wide range of management practices to meet specific environmental objectives. At Loddington, managed habitats have been developed to meet the requirements of specific wildlife species or groups of species, or to meet other environmental objectives. This includes the development of beetle banks which have been shown to reduce aphid pest densities to below threshold levels without the need for summer insecticides, as discussed in Chapter 5.

The EU Sustainable Use Directive and the UK's 25 Year Environment Plan promote Integrated Pest Management as a means of minimising pesticide use and making the greatest possible use of alternative measures, including natural predators of crop pests. Gaba and Bretagnolle (2020) found that numbers of insect pest predators and pollinating insects were influenced by the proportion of seminatural habitat in a French agricultural landscape. Redhead et al. (2020) report that at the landscape scale, seminatural habitats have an overall positive effect on wheat yield stability and resilience and attribute this to the presence of invertebrate predators, parasites and parasitoids of crop pests and pathogens. In this case, seminatural habitats represent land sparing at the field or farm scale but land sharing at the landscape scale, and an approach that is equally applicable to organic or conventional systems. Integration of small linear areas of seminatural habitat such as those adopted at Loddington into the landscape was also identified by Holland et al. (2020) as improving natural control of crop pests. However, in this pan-European study, an important overriding finding was that this was not consistent across study sites. This emphasises the need for local knowledge of the ecology of crop pests and their predators.

As discussed in Chapter 3, direct drilling is another approach that can be incorporated into conventional farming systems. Farmers often fall into either of the two camps – plough-based or direct drilling systems, with the strictest adherents to the latter maintaining that any soil disturbance is to be avoided. There are strongly held views on the relative merits of direct drilling of crops and plough-based systems. Many years of transition and adoption of direct drilling at the Allerton Project have taught us that the purist approach in favour of one or other option is not necessarily most advantageous, at least on clay soils. While some direct-drilled fields have performed as well as more conventionally managed fields, and at lower cost, other direct-drilled fields have been associated with problems such as slugs, grass weeds, compaction and waterlogging, necessitating occasional ploughing or subsoiling. As described in Chapter 3, practices that do not invert the soil, such as subsoiling, can be beneficial in difficult conditions or soil types.

A combination of direct drilling and cover crops is intended to improve soil structure, biology and function, with resulting benefits for crop production, biodiversity, water quality and carbon sequestration. The approach, especially when combined with grass leys and livestock in arable systems, is increasingly referred to as 'Regenerative Agriculture'. However, this term has been used since the 1980s to represent a very wide range of approaches to agricultural sustainability at a range of scales, including mulching and permaculture (Newton et al., 2020). Regenerative Agriculture is therefore a plastic term, and Newton et al. (2020) advocate that users of the term should 'define it comprehensively for their purpose and context'. This approach is adopted within the farming community and serves to facilitate a direction of travel towards more sustainable farming methods, even in the absence of a widely recognised definition.

Similarly, we found that many farmers interviewed as part of the Sustainable Intensification research Platform adopted practices that contributed towards sustainable intensification without fully understanding the term (Morris et al., 2017). In fact, 'Sustainable Intensification' is an apparent oxymoron which has proved to be alienating amongst some farmers. Terms such as this represent a banner to identify with and attract a following within the farming community, and amongst policy makers and others, but also detractors, largely because of their ambiguity (Newton et al., 2020). Both ambiguous and strongly binary terminology can therefore constrain adoption of policy and practice, and the language adopted can have a bearing on the level of integration and flexibility within farming systems at farm and national scales.

Dichotomies in food and water

Results from our catchment scale research have demonstrated that both domestic and agricultural sources of phosphorus and other nutrients influence water quality and ecology, as described in Chapter 6. These results have helped to break down the view that was once widely held that agricultural impacts, rather than discharges from the wider community, dominated the elevated phosphorus concentrations reported for watercourses across the UK and much of continental Europe.

Phosphorus discharges from small rural sewage treatment works in lowland England are not monitored or regulated and the cost of installing P-stripping technology at such a small scale is prohibitive for the water companies that are responsible for water treatment and supply in each region. Phosphorus concentrations in discharged water are therefore largely determined by the concentrations in sewage entering the works, and ultimately by the diet of residents within rural villages.

Approximately 20–30% of the dietary organic phosphorus for most adults comes from dairy products, another 20–30% from red meat, poultry and fish, and plant foods that are high in protein such as legumes, nuts and whole grains (Calvo & Uribarri, 2013). Additional inorganic phosphorus is derived from medicines and dietary supplements, and ironically, water, as phosphate is added to drinking water supplies to reduce corrosion of pipes. Dicalcium phosphate and pyrophosphate are examples of additives to food to improve texture, taste, appearance, shelf life and other food properties, especially in convenience foods such as baked goods, cheese products and processed meats (Calvo & Uribarri, 2013; Carrigan et al., 2014). These phosphorus additives account for 10–50% of adult total phosphorus intake, depending on individual diet.

Forber et al. (2020) found that the contribution of processed foods to total P consumption has increased in the UK between 1942 and 2016 from 21% to 52% (Figure 7.18), but consumption of total animal products has not changed significantly between these two reference years. Total P consumed in meat products increased 28% from 1942 to 2016, most likely due to the increased consumption of processed meats. Scenario analysis indicated that if individuals adopted a vegan diet or a low-meat ('EAT-Lancet') diet by 2050, the P burden entering sewage treatment works would increase

Figure 7.18 Supermarket shelves of processed food

by 17% and 35%, respectively, relative to the baseline diet. A much lower P burden increase (6%) was obtained with a 'flexitarian' diet which featured more meat, but not on a daily basis. These increases are attributed largely to the increase in consumption of whole grains, legumes and nuts which have higher P content per unit of protein than animal products (Forber et al., 2020).

There is some evidence that excessive phosphorus intake can disrupt hormonal regulation of phosphorus, calcium and vitamin D, contributing to impaired kidney function, especially in individuals who have existing kidney diseases (Lou-Arnal et al., 2014). This is consistent with other concerns about excessive consumption of some of the foodstuffs that are associated with high phosphorous concentrations, and the association with high-protein animal products establishes a link with vegetarian and vegan agendas. Diet therefore becomes politicised and linked to the development of personal identity, for example as a moral vegetarian (Chuck et al., 2016; Dimitrievska, 2014). Vegetarians and meat eaters represent another established dichotomy associated with individual identity. However, the identity of individuals can only be understood in relation to our interactions with others, and dietary choice needs to be considered in terms of households and wider social and cultural communities and practices as these can dominate the attitudes and values of individuals (Fox & Ward, 2006; Stajcic, 2013).

While elevated concentrations of phosphorus in stream water can legitimately be considered as either the responsibility of the individual consumer through their choice of food, or the responsibility of water companies through the economic decisions they make about water treatment in rural areas, in fact, community level, social and political values attached to diet, business and environmental strategies also play a major role. Far from being caused by either agriculture or domestic sources, elevated phosphorus in water is a complex societal issue involving individuals and the whole food system from production to supply. As with the other apparently binary issues discussed in this chapter, accepting this complexity and shared responsibility, is important to the planning of future change.

Note

1 *Berosus signaticollis, Dryops similaris, Haliplus laminatus, Helophorus granularis, Limnebius nitidus* and *Rhantus suturalis.*

References

Bessi, A., Coletto, M., Davidescu, G. A., Scala, A. & Quattrociochi, W. (2015) Science vs conspiracy: collective narratives in the age of misinformation. *PLoS ONE* 10 (2) e0118093. DOI:10.1371/journal.pone.0118093

BTO (2019) *The Breeding Bird Survey 2019.* British Trust for Ornithology, Thetford.

Carver, S. & Convery, I. (2021) Rewilding: time to get down off the fence? *British Wildlife* 32 (4), 246–255.

Chadwick, D. R., Cardenas, L. M., Dhanoa, M. S., Donovan, N., Misselbrook, T., Williams, J. R., Thorman, R. E., McGeough, K. L., Watson, C. J., Bell, M., Anthony, S. G. & Rees, R. M. (2018) The contribution of cattle urine and dung to nitrous oxide emissions: Quantification of country specific emission factors and implications for national inventories. *Science of the Total Environment* 635, 607–617. DOI:10.1016/j.scitotenv.2018.04.152

Chuck, C., et al. (2016). Awakening to the politics of food: Politicized diet as social identity. *Appetite* 107, 425–436.

D'Antonio, W. V., Touch, S. A. & Baker, J. R. (2013) *Religion, Politics, and Polarization: How Religiopolitical Conflict is Changing Congress and American Democracy.* Rowman & Littlefield, New York.

Defra (2019) *Wild Bird Populations in the UK, 1970–2019.* Department for Environment, Food and Rural Affairs, London.

Dimitrievska, I. (2014). Phenomenological investigation in the process of becoming a moral vegetarian. *Fork to Farm: International Journal of Innovative Research and Practice* 1 (1).

Douglas, K. M., Uscinski, J. E., Sutton, R. M., Cichocka, A., Nefes, T., Ang, C. S. & Deravi, F. (2019) Understanding conspiracy theories. *Advances in Political Psychology* 40 (1). DOI:10.1111/pops.12568

Dunn, J. C., Gruar, D. Stoate, C., Szczur, J. & Peach, W. (2016) Can hedgerow management mitigate the impacts of predation on songbird nest survival? *The Journal of Environmental Management* 184, 535–544.

Dutton, K. (2020) *Black and White Thinking: the Burden of a Binary Brain in a Complex World.* Farrar, Straus and Giroux, New York.

Elsheikha, H. (2021) Unpublished data.

Forber, K. J., Rothwell, S. A., Metson, G. S., Jarvie, H. P. & Withers, P. J. A. (2020) Plant-based diets add to the wastewater phosphorus burden. *Environmental Research Letters* 15, 094018. DOI:10.1088/1748-9326/ab9271

Ford, A. T., Ali, A. H., Colla, S. R., Cooke, S. J., Lamb, C. T., Pittman, J., Shiffman, D. S. & Singh, N. J. (2021) Understanding and avoiding misplaced efforts in conservation. *FACETS* 6, 252–271. DOI:10.1139/facets-2020–0058

Fox, N. & Ward, K. (2006). Health identities: from expert patient to resisting consumer. *Health (London)* 10(4), 461–479. DOI:10.1177/1363459306067314

Fuller, R. & Gilroy, J. (2021) Rewilding and intervention: complementary philosophies for nature conservation in Britain. *British Wildlife* 32 (4) 258–267.

Gaba, S. & Bretagnolle, V. (2020) Designing Multifunctional and Resilient Agricultural Landscapes: Lessons from Long-Term Monitoring of Biodiversity and Land Use. In: Hurford, C., Wilson, P. & Storkey, J. (Eds.) *The Changing Status of Arable Habitats in Europe.* Springer Nature, Cham. 203–224.

Grass, I., Loos, J., Baensch, S., Batáry, P., Librán-Empid, F., Ficiyan, A., Klaus, F., Riechers, M., Rose, J., Tiede, J., Udy, K., Westphal, C., Wurz, A. & Tscharntke, T. (2019) Land sharing/sparing connectivity for ecosystem services and biodiversity conservation. *People Nature* 1 (2) 262–272. DOI:10.1002/pan3.21

He, Y., Jones, P. & Rayment, M. (2017) A simple parameterisation of windbreak effects on wind speed reduction and resulting thermal benefits to sheep. *Agricultural & Forest Meteorology* 239, 96–107. DOI:10.1016/j.agrformet.2017.02.032

Holland, J. M., Jeanneret, P., Moonen, A-C., van der Werf, W., Rossing, W., Antichi, D., Entling, M. H., Giffard, B., Helsen, H., Szalai, M., Rega, C., Gibert, C. &

Veromann, E. (2020) Approaches to identify the value of seminatural habitats for conservation biological control. *Insects* 11 (3) 195. DOI:10.3390/insects11030195

Hoodless, A. (2020) Unpublished data.

Hoodless, A. (2021) Unpublished data.

Johnston, A., Robinson, R. A., Gargallo, G., Julliard, R., van der Jeugd, H. & Baillie, R. (2016) survival of Afro-Palaearctic passerine migrants in western Europe and the impacts of seasonal weather variables. *Ibis* 158 (3) 465–480.

Kendall, N. (2017) Unpublished data.

Kendall, N. R., Smith, J., Whistance, L. K., Stergiadis, S., Stoate, C., Chesshire, H. & Smith, A. R. (2021) Trace element composition of tree fodder and potential nutritional use for livestock. *Livestock Science* 250, 104560. DOI:10.1016/j.livsci.2021.104560

Kopnina, H., Leadbeatter, S. & Cryer, P. (2019) Learning to rewild: Examining the failed case of the Dutch "New Wilderness" Oostvaardersplassen. *International Journal of Wilderness* 25 (3). 72–89.

Lichtenberg, E. M., Kennedy, C. M., Kremen, C., Batary, P., Berendse, F., Bommarco, R., et al. (2017) A global synthesis of the effects of diversified farming systems on arthropod diversity within fields and across agricultural landscapes. *Global Change Biology* 23, 4946–4957. DOI:10.1111/gcb.13714

Lee, M., Stoate, C., Crotty, F., Kendall, N., Rivero, J., Williams, P., Chadwick, D., Morris, N., Clarke, D., Knight, S., Butler, G., Takahashi, T., McAuliffe, G. & Orr, R. (2017) *SIP Project 1: Integrated Farm Management for Improved Economic, Environmental and Social Performance (LM0201). Work Package 1.2B: Experimentally Test Innovative Practices and Technologies on Study Farms for Sustainable Intensive Farming.* Defra, London.

MacDonald, D. W., Newman, C. & Buesching, C. (2015) Badgers in the Rural Landscape – Conservation Paragon or Farmland Pariah? Lessons from the Wytham Badger Project. In: MacDonald, D. W. & Feber, R. E. (Eds.) *Wildlife Conservation on Farmland. Volume 2 – Conflict in the Countryside.* 65–95. Oxford University Press, Oxford.

Mieda, T., Taku, K. & Oshio, A. (2020) Dichotomous thinking and cognitive ability. *Personality and Individual Differences.* DOI:10.1016/j.paid.2020.110008

Morris, C., Jarrett, J., Lobley, M. & Wheeler, R. (2017). *Baseline Farm Survey – Final Report. Report for Defra project LM0302 Sustainable Intensification Research Platform Project 2: Opportunities and Risks for Farming and the Environment at Landscape Scales.* Defra, London.

Morrison, C. A., Robinson, R. A., Clark, J. A., Marca, A. D., Newton, J. & Gill, J. A. (2013) Using stable isotopes to link breeding population trends to winter ecology in willow warblers, *Phylloscopus trochilus. Bird Study* 60 (2) DOI:10.1080/00063657.2013.767773

Newton, P., Civita, N., Frankel-Goldwater, L., Bartel, K. & Johns, C. (2020) What is regenerative agriculture? A review of scholar and practitioner definitions based on processes and outcomes. *Frontiers in Sustainable Food Systems* 4, 577723. DOI:10.3389/fsufs.2020.577723

Oshio, A. (2012) An all-or-nothing thinking turns into darkness: relations between dichotomous thinking and personality disorders. *Japanese Psychological Research* 54 (4) 424–429. DOI:10.1111/j.1468-5884.2012.00515.x

Pettorelli, N., Barlow, J., Stephens, P. A., Durant, S. M., Connor, B., Buhne, H. S., Sandom, C. J., Wentworth, J. & du Toit, J. T. (2018) Making rewilding fit for policy. *Journal of Applied Ecology* 55, 1114–1125.

Redhead, J. W., Oliver, T. H., Woodcock, B. A. & Pywell, R. F. (2020) The influence of landscape configuration on crop yield resilience. *Journal of Applied Ecology* 57 (11), 2180–2190. DOI:10.1111/1365-2664.13722

Robinson, R. A., Lawson, B., Toms, M. P., Peck, K. M., Kirkwood, J. K., et al. (2010) Emerging infectious disease leads to rapid population declines of common British birds. *PLOS ONE* 5 (8) e12215. DOI:10.1371/journal.pone.0012215. Abstract

Pringle, H., Wilson, M., Calladine, J. & Siriwardena, G. (2019) Associations between gamebird releases and generalist predators. *Journal of Applied Ecology* 56, 2102–2113. DOI:10.1111/1365-2664.13451

Rose, D., Dicks, L., Sutherland, W., Parker, C., Lobley, M. & Twinning, S. (2016) *SIP Project 1: Integrated Farm Management for Improved Economic, Environmental and Social Performance (LM0201). Work Package 1.3A: Identifying the Characteristics of Effective Decision Support and Guidance Systems in the Context of Integrated Farm Management.* Defra, London.

Sage, R. (2020) The breeding success of hen pheasants. *GWCT Review of 2019.* 51, 46–47.

Scheelbeek, P., Green, R. et al. (2020) Health impacts and environmental footprints of diets that meet the Eatwell Guide recommendations: analyses of multiple UK studies, *BMJ Open* e037554. DOI:10.1136/bmjopen-2020–037554

Stajcic, N. (2013). Understanding culture: food as a means of communication. hemispheres. *Studies on Cultures and Societies* 28, 77–87.

Stoate, C., Whitfield, M., Williams, P., Szczur, J. & Driver, K. (2007) Multifunctional benefits of an agri-environment scheme option: riparian buffer strip pools within 'Arable Reversion'. *Aspects of Applied Biology* 81, 221–226.

Stoate, C., Morris, R. M. & Wilson, J. D. (2001) Cultural ecology of Whitethroat (*Sylvia communis*) habitat management by farmers: trees and shrubs in Senegambia in winter. *Journal of Environmental Management* 62, 343–356.

Stoate, C., Fox, G., Bussell, J. & Kendall, N. R. (2021) A Role for Agroforestry in Reducing Ammonia and Greenhouse Gas Emissions from Ruminant Livestock Systems. In: Aubertöt, Bertelsen et al. (eds) *Intercropping for Sustainability: Research Developments and Their Application. Aspects of Applied Biology.* 146. Association of Applied Biologists, Warwick.

Stoate, C. & Szczur, J. (2001a) Could game management have a role in the conservation of farmland passerines? A case study from a Leicestershire farm. *Bird Study* 48, 279–292.

Stoate, C. & Szczur, J. (2001b) Whitethroat *Sylvia communis* and Yellowhammer *Emberiza citrinella* nesting success and breeding distribution in relation to field boundary vegetation. *Bird Study* 48, 229–235.

Stoate, C. (1998) Abundance of Olivaceous Warblers (*Hippolais pallida*) and potential invertebrate prey in unmanaged *Acacia* woodland. *Bird Study* 45, 251–253.

Stoate, C. (1997) Abundance of Whitethroats *Sylvia communis* and potential invertebrate prey in two Sahelian sylvi-agricultural habitats. *Malimbus* 19, 7–11.

Stoate, C. & Moreby, S. J. (1995) Premigratory diet of trans-Saharan migrant passerines in the western Sahel. *Bird Study* 42, 101–106.

Stoate, C. (1995) The impact of Desert Locust *Schistocerca gregaria* swarms on pre-migratory fattening of Whitethroats *Sylvia communis* in the western Sahel. *Ibis* 137, 420–422.

Stoate, C. & Szczur, J. (2006) Potential influence of habitat and predation on the recovery of a Biodiversity Action Plan species, the Spotted Flycatcher *Musciapa striata*. *Bird Study* 53, 328–330.

Vickery, J. A., Ewing, S. R., Smith, K. W., Pain, D. J., Bairlein, F., Skorpilova, J. & Gregory, R. (2013) The decline of Afro-Palaearctic migrants and an assessment of potential causes. *Ibis* 156 (1) 1–22.

Walker, B. (2019) *Can supplementary willow fodder increase blood cobalt and vitamin B_{12} concentrations in weaned lambs?* Year 3 research project dissertation. University of Nottingham School of Veterinary Medicine and Science.

8 The view from the farm

The policy context

As in the rest of Europe and many other countries, the way farmers manage their land in the UK has been influenced considerably by a high-level policy context, specifically in our case, the European Union's Common Agricultural Policy (CAP). Initially, economic and policy support for farming was linked to output, but subsequent measures to curb excessive production included an obligation to take land out of productive use in the form of 'set-aside'. This fallow land was becoming a feature of the arable landscape when the Allerton Project started in 1992.

Subsequent reform of the CAP in 2005 included modulation (diverting funds from the direct support to payments for rural development and environment) and decoupling of subsidies from production, the introduction of cross-compliance measures for observing basic environmental standards, and extension of the Rural Development Programme, including further development of the Agri-Environment Programme. Later, obligatory modulation rates increased, set-aside was abolished, changes were made to cross-compliance, and new requirements were introduced to retain some of the environmental benefits of set-aside and improvements to water management.

Under this policy, farmers received payments if they maintained their land in good agricultural condition and complied with standards on public health, animal and plant health, the environment and animal welfare (EU Council Regulation no. 1782, 2003). Under the Rural Development Programme, agri-environment schemes have aimed to enhance the ecological status of farmland through land management targeted mainly at biodiversity conservation and improvement in water quality.

The structure of the Agr-Environment Programme is pyramidal. Its base is formed by 'horizontal' schemes that apply to all agricultural land and provide support for environmentally friendly production methods. Higher on the pyramid are area-specific zone schemes that target High Nature Value areas and focus on nature conservation and landscape protection. Member states decide on the national priorities, with many countries focusing

DOI: 10.4324/9781003160137-8

on environmental issues and resource protection and a varying priority towards biodiversity conservation. The agri-environment programme has had clear economic impacts in the UK, representing up to 40% of farmers' income, but farmer participation in agri-environment schemes is influenced by a wide range of socio-economic and cultural factors.

A series of evolving agri-environment schemes has enabled many farmers to undertake additional management to deliver more targeted environmental benefits, especially in terms of wildlife conservation. Initially, the Countryside Stewardship scheme enabled farmers with an interest in wildlife conservation to create and manage habitats. From 2005, a similar Higher Level Stewardship scheme ran alongside an Entry Level Stewardship scheme which was designed to encourage those with more production-oriented approaches to adopt similar management at a lower level of commitment.

Following the UK's departure from the EU in 2020, UK agricultural and agri-environmental policy entered a transition period in which area payments to farmers were reduced (ultimately to zero in 2028) and a new Environmental Land Management scheme has been in development, based on a principle of 'public money for public goods'. Such an approach is also being considered for the EU, and the UK therefore represents a testbed for policy that could be adapted and adopted across EU member states and other countries outside the EU.

The UK approach takes the form of three schemes that reward environmental goods: the Sustainable Farming Incentive (SFI), the Local Nature Recovery scheme, and the Landscape Recovery scheme. All are intended to deliver 'clean and plentiful water', 'clean air', 'thriving plants and wildlife', 'reduction in and protection from environmental hazards', 'adaptation to and mitigation of climate change', and 'beauty, heritage and engagement with the environment' (Defra, 2018). The SFI is the most widely accessible approach, but Local Nature Recovery schemes will be adopted in areas where collaboration between farmers to meet landscape scale wildlife conservation objectives would be feasible and beneficial.

Other EU policies have had a considerable impact on how land is managed, and continue to do so, despite the UK's departure from the EU as the UK government committed not to erode existing standards. Together with the Groundwater Directive and the Nitrates Directive, the Water Framework Directive provides a model for a system of water management by river basin and gives high priority to ecological status. Good ecological status is defined in terms of the quality of the biological community, hydrological and chemical characteristics. There is also a high level of protection for groundwater and the Directives require designation of Nitrate Vulnerable Zones and modification to farming practices to protect water through Codes of Good Agricultural Practice.

At and around the Allerton Project, farmers have therefore changed their land management to comply with these policy changes. The area is within a Nitrate Vulnerable Zone and there are constraints on the use of nitrogen fertilisers with strict requirements for application according to each crop's requirements and documentation of the evidence base for this. Limits on pesticide concentrations in drinking water supply place restrictions on how pesticides can be used on farmland, and the regional water company is active in encouraging farmers to reduce the use of the most problematic pesticides in terms of drinking water treatment and reducing movement of pesticides from agricultural land to water. Over the past 30 years, there has therefore been an increase in the emphasis on land management to improve water quality, alongside water treatment to remove pollutants from drinking water supply.

Whereas in 1992, it was commonplace for broad-spectrum herbicides to be applied to field margin hedge bases so that herbaceous vegetation was destroyed, there has more recently been a requirement to leave a minimum 2-metre strip of unsprayed perennial vegetation at the base of the hedge, as measured from the centre of the hedge. There are also strict limits on the use of pesticides on the productive area in field headlands to minimise spray drift and surface runoff to field boundary habitats and watercourses.

In addition, the number of plant protection products available to farmers has declined over the past 30 years as those with the greatest demonstrable negative impacts on the environment or human health have been withdrawn or restrictions placed on their use. The risk of exceeding the 0.1 μg/L limit for drinking water supply is also an influence on product availability for use on farmland, independently of impacts on human health or the environment. Restrictions on pesticide use can have an influence on the crops grown at the farm and landscape scale. For example, withdrawal of neonicotinoid insecticides in 2018, resulted in a substantial increase in damage to oilseed rape crops by cabbage stem flea beetle *Psylliodes chrysocephala*, and subsequent reduction in the area grown.

The global market

A major driver of crop selection by farm businesses is the global price of agricultural products, including arable crops, meat and dairy. These are the main products from farmland across much of lowland England, including the area around Loddington, and they are sold from the farm primarily into a global market at a price that is determined by global supply and demand. As such, local farmers have only limited control over their income from sales.

Global prices of cereal crops rose significantly in the 2006–2008 period, triggering a food crisis in some countries (Headey & Fan, 2010). This was primarily driven by a combination of rising oil prices, a greater demand for biofuels, and trade shocks in the food market. Oil prices affect grain prices, mainly directly through their impact on input costs such as diesel, machinery and fertiliser, but also indirectly by increasing demand for biofuels, reducing the area of food grains grown. Higher energy prices increase the demand for biofuels which become more competitively priced when compared with oil. This process drove up the demand for biofuels derived from maize in the United States in the 2006–2008 period for example. The problem was compounded in some countries by export restrictions, especially of rice, and panic buying of rice and other grains. Droughts in Europe, North America and Australia from 2006 to 2008 put additional pressure on grain prices, and Ukraine and other major cereal exporting countries restricted or banned exports because of poor harvests. The result was a doubling of global wheat prices.

Further, albeit lesser increases in price occurred in 2010–2012, and in 2020, droughts in South America were largely responsible for another increase. Concerns about food security and related risk of political instability also resulted in some Asian and African countries increasing imports. With the exception of a peak during the fuel crisis in the early 1970s, global food prices reached their highest level since 1961 in real terms in 2021 (FAO data). Most recently, Russia's invasion of Ukraine in February 2022, resulted in immediate substantial increases in the prices of cereals, vegetable oils, fuel and fertiliser. The combined and related threats of drought-induced low crop yields, food insecurity and political instability are increasing concerns at the national and global scale, but also have impacts for local producers through changes in crop prices and input costs that are beyond their control. There are associated implications for both the profitability of farm businesses and the extent to which environmental management is carried out within and outside the productive area.

Business strategies

The extent to which farms can adapt to changes in global commodity prices and national or European policy is influenced by their business type. In 2015, an interview survey of 34 farmers was carried out in the upper Welland as part of Defra's Sustainable Intensification research Platform (SIP, Morris et al., 2017). The term 'sustainable intensification' was used to describe the process of improving or maintaining productivity while simultaneously maintaining or improving the environment. It includes management practices which exploit synergies between the two objectives such as improved crop or livestock genetics and nutrition, nutrient use efficiency, and optimising the use of marginal land.

Seven of the farms surveyed were less than 100 ha in area, 21 were between 100 and 500 ha, and 6 were over 500 ha. The work was part of a larger study in which a total of 244 farmers were interviewed across seven English and Welsh study areas. The majority of the farmers interviewed in the Welland (28 of the 34) used contractors for some of the operations on the farm. On average, the farm business represented 60% of household income, with the rest being made up of a range of on-farm and off-farm activities, and income from investments. Twenty of the 34 Welland farmers reported their business as being in a poor or fair economic position, and none classified their economic position as 'excellent'. Eighteen of the farmers claimed to be satisfied with their level of production at the time, and eleven were less than satisfied.

At the time, farmers reported low global commodity prices, high input costs and uncertainty over future agricultural policy as constraints on both profitability and planning for the future.

26% of the farmers claimed to have made no significant changes to their businesses in the previous five years. Amongst those that had, 24% had made changes to farm practices, 15% had reduced the size of the business, and 12% had expanded it. Welland farmers were less likely to have expanded their businesses than those in most of the other areas surveyed, and large farms were more likely to have invested in business expansion than small ones were.

Fifty-three percent of farmers planned no change to their business over the following three years, slightly higher than in the other regions where the survey was conducted. Fifteen percent were considering expansion of their existing business, while 9% were considering contraction. Investments were sometimes made out of necessity, rather than to expand the business, for example, the construction of a building to house cattle in the face of increasingly wet winters.

In a previous survey of 29 farmers in the Eye Brook catchment in 2011, 31% reported that they had experienced some impact of climate change on their farm businesses, mainly in the form of drier springs reducing crop growth. One respondent was considering drought resistance when selecting crop cultivars, and another was starting a carbon audit. Two of the farmers who had not noticed the impacts of climate change identified reduced tillage as a means of addressing future challenges, and another was adopting renewable energy generation.

The year 2012 also started with a dry spring but ended up being the wettest year on record for England, and the second wettest for the UK as a whole. Dry springs were often experienced through subsequent years, and exceptionally prolonged drought conditions occurred through 2018, but wetter autumns with more intense storm events through the autumn and winter also became more common. Rainfall from September through the autumn and early winter of 2019 was so severe that almost the entire arable area remained undrilled with crops and there was widespread flooding. There has been a tendency for increased soil compaction over recent years as the early autumn period has become wetter. This results from both grazing livestock and the use of increasingly heavy machinery for arable operations at this time of year. Climate change is therefore starting to influence the way farmers manage their businesses.

Linked to the SIP, Ang et al. (2016) analysed the economic and environmental performance of a sample of English arable farms, based on Farm Business Survey data for 2012 (when adverse weather affected yield) and

2013 (when the wet conditions affected crop establishment on many farms). In both years, the frequency distribution of overall sustainability efficiency is bimodal. In 2012, one group was more than 90% efficient, while the other was less than 80% efficient. In 2013, both groups became less efficient, but this was especially the case for the group that was already less efficient. The authors attribute this to the challenging conditions associated with clay soils such as those around Loddington. However, the results that relate specifically to nutrient use efficiency are different. Both groups are less efficient in 2013 than in 2012, but the normally more efficient farms are affected most, possibly because the nutrient requirement was over-estimated or optimum timing of application could not be achieved.

Land-use decisions

Given the unpredictability of global crop prices, increasing input costs, and challenges associated with climate change, the options available to farmers to optimise their economic return are limited. There is a trade-off between the need to reduce input costs and increase yields per unit area of land. Maximising yield does not necessarily improve economic performance if input costs are not sufficiently managed. This is illustrated by data from local arable farmers in 2017 (Figure 8.1).

Farms vary considerably in their approaches to managing input costs and there is no relationship between these input costs and crop yield. One local farmer grew crops for seed, obtaining a premium in terms of income, but other farms were selling into markets for which the price was determined globally.

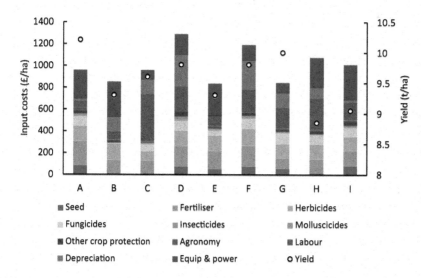

Figure 8.1 Winter wheat yield and input costs for upper Welland farms in 2017

I would say, across the board, it's a struggle at the moment because the prices are so low ... Most of the stuff that you sell, grain for instance is 75% what it was ... So, there's not a lot of money left to play with ... And when farmers are making plenty of money, they spend it, so it's not just farming that suffers.

(Arable farmer, Upper Welland)

Farms varied in the size of machinery they adopted and in the frequency with which machinery was replaced. Farmers had a range of approaches to soil management and crop establishment, from conventional use of the plough and subsequent passes to create a seedbed, through reduced tillage systems, to direct drilling into the stubble of previous crops. The labour, equipment and power costs decline across this spectrum. The lowest fertiliser cost was 56% lower than the highest, and the lowest pesticide cost was 46% lower than the highest. These costs will not only reflect purchase costs, often linked to global oil prices but are also a reflection of actual use on farm.

In a study linked to the SIP, Lynch et al. (2018) used data from the Farm Business Survey (FBS), an extensive annual economic survey of representative farms in England and Wales, to explore the relationship between these economic data and environmental indicators derived from the survey returns. They found that, for arable farms in the East of England, profitability was associated with nitrogen use efficiency (based on theoretical requirements of the crop) and greenhouse gas emissions efficiency. There was no such relationship for phosphorus use because of the complex on-farm processes associated with this nutrient, as described in Chapters 3, 6 and 7. This study also worked with data from dairy farms in the southwest of England and found no significant associations between profitability and environmental externalities.

For many years, winter wheat and oilseed rape have provided the greatest economic return on investment in inputs, and have therefore been the main crops grown, but the input costs are also the highest, making them more risky in drought or excessively wet conditions. Other crops have lower input costs but also lower income. Despite this, and the dominance of a simple wheat/rape rotation on many farms for many years, these other crops are also incorporated into the rotation as a means of controlling pests or diseases, or building fertility. For example, as a legume, field beans require minimal fertiliser (84% lower cost than wheat in the current example) and contribute nitrogen to the following crop in the rotation, while also providing an opportunity to use herbicides that control grass weeds such as black-grass. Vigorous cereal crops such as autumn-sown oats and modern hybrid barleys provide a break from fungal diseases of wheat and compete well with black-grass, with lower fertiliser and pesticide costs. The data from local farms for 2017 reveal that fertiliser costs for oats were half those of wheat, and pesticide costs were 60% lower (Figure 8.2). Pesticide costs for

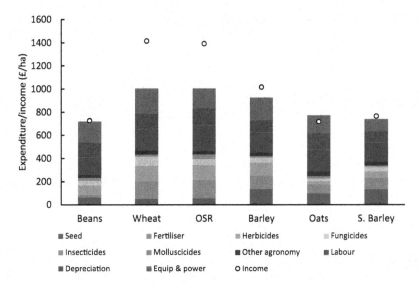

Figure 8.2 Input costs and income from winter wheat compared with oilseed rape and other break crops on local farms in 2017

oilseed rape were only marginally lower than those for wheat, while pesticide costs for winter barley, winter beans and spring barley were respectively 26%, 35% and 50% lower than for winter wheat. While the economic return on spring-sown crops is low, they provide an opportunity to control difficult weeds using stale seedbeds and a broad-spectrum herbicide in the autumn.

In the 2015 SIP survey, one arable farmer reported:

> I was leaning on wheat/rape as profitable, but I moved out of that as I realised it's not sustainable – so I'm now growing a wider rotation. I realised that you can't keep doing that and expect optimum yields, and you risk the problem with black-grass. And I think this idea of having a legume in the rotation is right anyway. We've got to go back to a rotation to improve the organic matter.

Farmers involved in the Water Friendly Farming project described in Chapter 6 identified similar constraints (Stoate et al., 2019). They mentioned a tension between the economic pressure to adopt a short relatively profitable crop rotation of wheat and rape and the longer-term costs of this approach in terms of weed and disease control. The short rotation also reduces the diversity of habitat available to wildlife and increases the application of individual pesticides at the catchment scale, but there is a short-term risk of adopting a longer more diverse rotation with a smaller area of high return crops.

But when you think of the production cycle for farming livestock, is probably a three-year production cycle before it comes in. We're planning crops now, which won't be marketed for two years and we're doing this rather blindly, not knowing what's going to happen in two or three-years-time. But it's all very relevant to the here and now.

(Water Friendly Farming project arable farmer)

As part of the SIP, Ang et al. (2016) explored the economic efficiency of crop diversification, based on FBS data from a small sample of 44 English arable farms for the 2007–2013 period. They found that 67% of farms had a negative opportunity cost for crop diversification (i.e. they would gain by increasing crop diversity), while 15% had a positive opportunity cost. However, the standard deviations were large, highlighting the substantial variation between farm businesses.

The arable rotation within the Water Friendly Farming project study area and other local farms has been dominated by winter wheat, with oilseed rape as the dominant break crop until the ban on the use of neonicotinoid insecticides made rape more risky and less profitable. However, winter beans, winter oats, and spring barley have all also been incorporated into the rotation for the reasons given above. Wet conditions in the autumn of 2019 resulted in a predominance of unprofitable spring-sown crops in 2020, and the abandonment of rape from the rotation resulted in a much larger range of alternative break crops being grown in the following year (Figure 8.3).

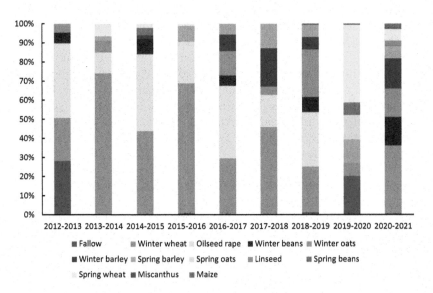

Figure 8.3 Arable crop areas in the Eye Brook and Stonton Brook headwaters by year

The wet year of 2012 also resulted in the presence of fallow land, and in very low yields of those crops grown. At Loddington, winter wheat yields were 40% below the ten-year average (Figure 8.4). The exceptional drought conditions of 2018 did not influence the cropping or crop area as crops were already drilled before the drought started, but yields were reduced. Compared with the previous long-term average, excluding 2012, winter wheat yields at Loddington in 2018 were 10% lower. As in the wider landscape, the wet autumn and winter of 2019 almost completely precluded the drilling of winter wheat and forced farmers into spring drilling of less profitable crops or undrilled fallow land. Only spring-sown wheat was grown in that year at Loddington, resulting in low overall production and a 40% reduction in yield on those fields where it was grown. Even excluding these extreme values, food production foregone as a result of taking land out of production to meet environmental objectives is considerably lower than the between-year fluctuations in production associated with variable yields (Figure 8.4) and provides a potential means of reducing economic risk if payments are sufficient and environmental measures are carefully targeted at the least productive land.

As well as the diversity of crops, their dispersion within the farm, and across the landscape, influence their accessibility to wildlife species. However, farms have adopted 'block cropping' of several fields together in the same crop for good economic reasons. At Loddington, crops were, at one stage distributed across the farm to improve their availability to foraging birds, but this was subsequently abandoned because of the additional costs

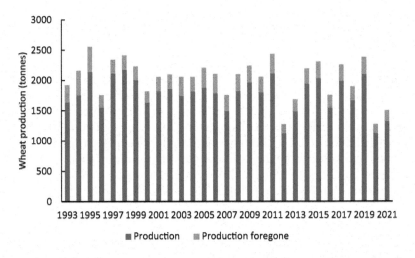

Figure 8.4 Arable crop production at Loddington using wheat as a proxy for all arable crops, and yield and arable area for each year. Production foregone represents the amount of crop that would have been harvested had the land not been taken out of production to meet environmental objectives

associated with dispersing all field operations for each crop over a wider area, and because alternative habitats had been integrated and dispersed into the cropped landscape to meet the ecological role of foraging habitats for birds and other wildlife.

Land management decisions

Members of the local farmer network identified a lack of knowledge about soil organic matter, and declining levels in soil, as a key factor influencing the sub-optimal profitability of their businesses. Soil organic matter is important for crop rooting capacity and nutrient uptake, for trafficability and moisture retention, but it is also important for delivering societal benefits such as improved water quality and reduced flood risk through its influence on infiltration and runoff (see Chapter 3). Similarly, farmers in the Water Friendly Farming project (Stoate et al., 2019) identified soil compaction as an increasing constraint associated with increasing autumn rainfall and farm machinery size, and this is also a major influence on infiltration and runoff. Soil management is therefore fundamental to achieving both agricultural and environmental objectives.

While soil organic matter and compaction are soil characteristics that are identified by farmers, scientists and policy makers, the language and labels attached to soil characteristics are more nuanced than they initially appear. Jones (2019) took a transdisciplinary approach to the study of soil quality using methods from both social and soil science disciplines, as well as drawing upon academic and non-academic knowledge. This study in the East Midlands combined semi-structured interviews and participatory exercises with analysis of soil that farmers had identified as 'good' or 'bad', and follow-up interviews with farmers to see how they interpreted the scientific assessments.

One of the key elements of the initial discussions was investigating what farmers' understanding of soil quality was and what they regarded as a "good" and a "bad" soil. This therefore provides another example of binary thinking discussed in Chapter 7. The issue was first explored using participatory graffiti wall exercises (Hanington, 2003) during two farmer workshops which took place at Loddington. Attendees were given "post-it" notes on which they could write their thoughts on what made a "good" or a "bad" soil. These concepts were further investigated during 20 semi-structured interviews with farmers participating in the study.

The context in which these words are employed can potentially affect farmers' responses. In the graffiti walls exercise, where the context encouraged farmers to consider *a soil* in an abstract sense, responses were largely dominated by what the scientific literature on soil quality would term *dynamic* soil quality indicators, aspects which change over time. The most prominent characteristics reported by the farmers related to soil structure and organic matter. However, in the interviews, where the farmers had been speaking about their own farms and were asked about *their* "good" and "bad" soil, farmers were more likely to discuss the *inherent* characteristics of their soil, especially its texture.

Characterising "good" and "bad" soil is therefore not simple. For example, in wet years, faster draining, "lighter" soils might be "good" because they allow the water to drain away and prevent waterlogging, but the same soil may be "bad" in a dry period because it will not retain the water, causing the crops to suffer from drought.

There is also the complication of how easy it is to work a soil versus how it yields. Farmers often spoke of how their better soils required less work to establish a seedbed. However, these "good soils" at the time of working are again the lighter ones which might not provide sufficient water to crops in a drought. In contrast, heavy land that might be difficult to manage to create a seedbed can subsequently provide a high yield in dry conditions because of good moisture retention.

There is a widespread recognition amongst farmers participating in research that the soil management approach currently being widely adopted is unsustainable and that there is a need for a different approach to be developed, promoted and adopted:

> it's such an exciting industry to be involved with, the potential for improvement is so great. The potential for causing damage is also so great and it's finding a nice balance between which way to go, and it comes back to possibly the education side of things. Getting people more involved and more research being required to understand where we've gone wrong.
>
> (Water Friendly Farming project arable farmer)

> For far too long, in my opinion, we've looked at the physics of the soil... we've looked at the chemistry of the soil through the fertiliser companies; nobody has given any consideration to the biology within the soil. They've looked at it from three different aspects completely, or two aspects ... the commercial aspect of the chemistry and the physics. The biology has been totally neglected ... and we've got to get back to having soil in equilibrium and balance.
>
> (Water Friendly Farming project arable farmer)

Land tenure is an important influence on the type of management adopted, with short-term tenancy arrangements being particularly restrictive of long-term plans for management and investment in long-term objectives for soil for example. From a contractor's point of view, soil management depends on what the owner wants to achieve. The contractor has no opportunity to move to or invest in a direct drilling approach if the landowner wants maximum yields and will get in another contractor to protect their short-term returns. However, where a farm is managed in hand, the owner can take a longer-term view. This concern was expressed by farmers participating in the Water Friendly Farming project, and those engaged in the SIP.

> Well, we're doing this strip tillage now and ... one block of land has done three years and another block is into two years. There's been a marked improvement in soil condition in the top six inches of soil. It's just stunningly, remarkably improvement. I agree with what you say, to go from what we've been doing, straight into it, a no till situation would be drastic. You've done the same at Loddington, you've gone into no till, almost and I don't know whether that's a long-term approach. But I think from soil management and improving it, that would be achievable after five, six, seven years of judicious soil management.
>
> (Water Friendly Farming project arable farmer)

> They are not going to invest as heavily in someone else's land as they would their own ... It's a completely different mindset ... You're not going to make the investments.
>
> (Farmer interviewed for the SIP)

> The basic requirement is the highest output possible. Because that is where that landowner receives his income from on a contract farm basis. ... You have to remember that these things actually only work on an owned in-house farm block. When you start talking contract farming then [as a contractor] you've got to go to an individual landowner and say ... how do we integrate your block of land into this? That doesn't interest me in any way, shape or form.
>
> (Water Friendly Farming project contractor)

> Although the vast majority of our farm is on, supposedly a short-term basis, you have to try your hardest not to view it in that respect. Because you don't necessarily want to give it up, I don't want to give anything up. Well, okay, I've got another five years, dammit, I wish I'd done that ... You have to try to but, you've also got to try and base that with the thought at the back of your mind that, you might not be there in five-years-time or in twenty-years-time.
>
> (Water Friendly Farming project tenant farmer)

> I guess with modern shorter-term tenancies, the FBT land, you're going to have a slightly shorter viewpoint on it because you're not building it up for the next generation. And again, with contract farming the contractor is looking to maximise the profit on the land and that's how he makes his return.
>
> (Water Friendly Farming project farmer)

Good husbandry evolves over a long, long time. Short term, which is driven by finance, doesn't and government policy doesn't look at a long-term approach.

(Water Friendly Farming project farmer)

Improving organic matter in soil should be something that is focussed on by Defra. Really good solid work done and financed by them, and not leave it to individuals to experiment. Come up with some blueprint that is paid for either by government, Defra or the water companies to make soil far more resilient.

(Water Friendly Farming project farmer)

Learning through agri-environment schemes

The increasing adoption of agri-environment schemes over the past 30 years has had a considerable influence on land management. By the time of the 2015 SIP survey (Morris et al., 2017), 26 of the 34 surveyed farms had an agri-environment scheme agreement, of which eight were at an advanced level such as Higher Level Stewardship, and 16 were at a standard level (e.g. Entry Level Stewardship). Larger farms were more likely to have agri-environment scheme agreements than small ones, both locally in the Welland, and across the other six study areas in England and Wales. Other structural character-istics of farms such as land use and area of semi-natural vegetation are also known to influence agri-environment scheme participation.

Involvement in agri-environment schemes also appears to influence some management practices beyond those within the agreement itself. In the 2015 SIP survey, there was an apparent association between farms' agri-environment scheme participation and their adoption of some management practices that fall within the wider 'sustainable intensification' umbrella (Figure 8.5). Across all the farms in the 2015 SIP survey, 99% of those in the more advanced agri-environment schemes claimed to be already optimising marginal land for wildlife, compared to 60% for those farms that did not have an agreement. There was also an indication that farms with advanced agreements were more likely to adopt some adaptive activities to build re-silience into their businesses. Fifty-eight percent of farms with an advanced agreement were adopting stress tolerant crop cultivars, compared to 29% of those without an agreement. Farms without agreements were also less likely to be adopting or considering precision farming techniques. Farmers with advanced agreements had a better understanding that sustainable intensifi-cation was about improving both production and environmental objectives than those without agreements.

As in the other study areas, hedge and field margin management were most often cited as the environmentally beneficial management practices, but reductions in pesticide and fertiliser use were not mentioned by any of

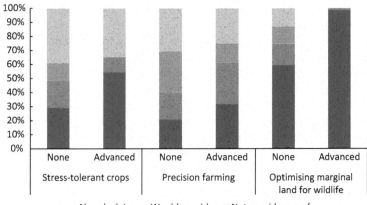

Figure 8.5 Adoption of sustainable intensification approaches by farmers with and without agri-environment scheme agreements (from data in Morris et al., 2017)

the farmers in the Welland, whereas they were in the other areas. However, 32% of the Welland farmers interviewed accepted that their farming had some 'detrimental environmental impact'. For example:

> Despite what we do in terms of margins and application, here we are on a sloping part of [the area], we must be getting run-off into water courses … It does concern me.
>
> Spraying is always the concern - but I can't see how it's really avoidable. We can be careful, when we spray with as far as insecticides go, bees and such like … It's something we can't really get round.
>
> But of course, farmers are trying to reduce inputs, trying to increase yields, trying to minimise their impact on soils and water. You know I think that's what we're all trying to do. But whether or not everyone can do the same thing and be successful is another matter. And I don't think they can. I think it's different for different farmers with different soil types. But I think that's the fundamental drive for most farmers.

The influence of involvement in agri-environment schemes on farmers' wider management, and attitudes to environmental issues also has an important temporal element, with farmers understanding and commitment to environmental management evolving through their participation in agri-environment schemes through time. This was explored in detail by Jarratt (2013) and Jarratt et al. (2013) and further unpublished analysis which applied the sociological concept of a 'moral career' across farms in the East of England. Associated with this conceptual framework are 'contingencies' or transformative moments in the 'career' of an individual, and 'agents', normally individuals who have some influence on changes in career pathways.

Table 8.1 Summary of Environmentally Friendly Farming careers and stages

EFF career	Career stage	Short form	Definition	No. of farmers at time of interview
Environmental Wage	Productivist	P	Exclusive focus on agricultural production	0
	Profit protection	PP	Agri-environment schemes supplement income and spread risk	5
	Basic conservation	BC	Participation in lower-level agri-environment schemes	3
	Advanced conservation	AC	Participation in higher-level agri-environment schemes	9
Environmental Opportunity	Unsupported	U	Conservation activities carried out on a voluntary basis	7
	Already doing it	ADI	Payments received for conservation activities already adopted. No specific conservation objectives	0
	EFF for wildlife	EW	Combination of financial support and wildlife interest provides an opportunity to implement conservation work	1
	EFF for species	ES	Combination of financial support and interest in specific species provides an opportunity to develop targeted conservation work	8
	EFF leader	EL	Sharing knowledge with wider farming community and encouraging landscape scale collaboration	10

Forty-three farmers were recruited to the research, comprising owners (26), tenants (7), managers (9) and an agent, 30 of the farms exceeding 200 ha in size. Importantly for this temporal study of environmental learning through participation in agri-environment schemes, there was representation from the very earliest Environmentally Sensitive Areas, initiated in 1987, to the most recent Entry Level Stewardship scheme.

The first finding of the research was that for most farmers engaged in on-farm environmental action, there is an evolution and development in this engagement over time and this can be meaningfully captured by and represented through the concept of an Environmentally Friendly Farming 'career'. It was possible to differentiate two Environmentally Friendly Farming careers: the 'Environmental Wage' and 'Environmental Opportunity' careers, both of which are made up of distinct career 'stages' (Table 8.1 and Figure 8.6). Each career is distinguished by characteristic 'moral' dimensions. In the case of the Environmental Wage career, the emphasis is on the

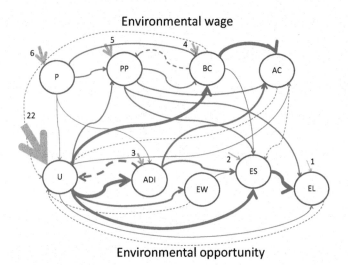

Figure 8.6 Conceptual model of farmers' environmental wage (top) and environmental opportunity career pathways (bottom). Orange arrows indicate the initial entry points, with adjacent numbers indicating the number of farmers involved. Blue arrows indicate the movement of farmers between careers and career stages, with line width indicating the number of farmers involved. Dotted blue arrows indicate backward movement through careers. Refer to Table 8.1 for abbreviations (Susanne Jarratt, 2021)

significance of financial incentives in prompting and delivering environmental actions, while in the case of the Environmental Opportunity career the emphasis is more on a desire to achieve environmental benefits for their own sake. Although each of the careers is a typical career, the career of an individual farmer can be non-linear so a farmer may leapfrog or repeat stages and an individual farmer may also move between careers.

A second key finding to emerge was the changes in pro-environmental actions that were captured by the career concept as revealing of 'progress' in those actions in the sense that farmers deepen their knowledge of, commitment to and confidence about the environment as they pass through the different stages of their Environmentally Friendly Farming career. This was observed in the context of both careers, although to a more profound degree in the Environmental Opportunity career. In association with these shifts, what farmers actually *do* in relation to the environment also changes and develops. For example, their efforts become more focused and targeted on the management of particular species, rather than the amorphous concept of 'wildlife' and this results in greater 'ownership' of individual species. Some farmers were enthusiastic about the conservation of stone curlews *Burhinus oedicnemus*, while for others, lapwings *Vanellus vanellus* or grey partridges *Perdix perdix* were the focus of management activities. While stone curlews have a very restricted range, each of these birds is a large, easily identifiable and iconic species, with grey partridge and lapwing being strongly associated with productive land use. Some farmers in our study took great pride in their conservation of these species, to the point of introducing a competitive element to wildlife conservation on farmland. The presence or abundance of these species became a symbol of social status. Farmers' strengthened ownership of environmental management also encouraged experimentation and development of modified management practices that work for them.

> We are aware that this escarpment ..., it always has been incredibly important for lapwings as well and we are now getting a lot of pleasure by seeing hundreds of them up there.
>
> I am of the opinion that pollen and nectar mixes shouldn't have any grasses in them at all, because all the grasses do eventually kill all the flowers and take over the flowers, which just reduces the longevity of your margins. We then went to just a pure - whether it was clover or whatever flowers they were - and they were much easier to manage. So that was something I did.

More recently, less charismatic species have caught the attention of farmers. Those adopting reduced or no-tillage systems of crop establishment have noticed increases in earthworm numbers and it is not unusual to hear farmers discussing amongst each other the number of 'casts' on the soil surface, or the number of earthworms per spade full of soil in different fields.

Development of commitment can therefore apply to the management of habitats or species. This type of transition has subsequently been observed in the context of the Game and Wildlife Conservation Trust's (GWCT) 'Big Farmland Bird Count' which has encouraged farmers with limited knowledge of bird identification to learn about and count birds on their own farms, potentially stimulating progression from the ADI or EW to ES career stages (see Table 8.1 and Figure 8.6). However, in addition to the observed tendency of Environmentally Friendly Farming careers to involve 'progress' (which implies a degree of linearity), the research also found that Environmentally Friendly Farming careers can be reversible. The movement that has been identified between the two Environmentally Friendly Farming careers, albeit occurring to a relatively limited extent 'within career' moves, is further evidence of the complexity of this type of moral career.

The third key finding of the research was that agri-environment schemes play a crucial role in the development of farmers' Environmentally Friendly Farming careers and that for many, the progression in their environmental activities would not have occurred had the state not encouraged and financially incentivised them to do so.

> You can't really call it 'financially viable', because, that's a very debatable point. But, until it had financial assistance, it probably wouldn't have happened

Agri-environment schemes drive farmers' progression through the different career stages as career contingencies are created by agri-environment schemes being rolled out, withdrawn and replaced with new schemes. Research has not previously provided the evidence that long-term involvement with these initiatives can have the 'developmental' effects that this study revealed. The impact of this intervention goes beyond the provision of financial support for environmental actions to include, most notably, the contribution of advisors from government and non-government organisations to farmers' career progression. Advisors can help create career contingencies which move farmers from one career stage to another and this is especially important for farmers when moving from a basic to a more advanced scheme.

> [*A Natural England project officer*] who had obviously done this over many farms, told me precisely how to get the floristic margins in for instance, which was not drilling but just scattering and it works superbly, my floristic margins are superb.

The research also suggested that there is a 'legacy' effect of agri-environment schemes when farmers discontinue their participation but maintain some of their environmental actions.

A further finding relates to the identification of the final stage of the Environmental Opportunity career. Environmental leaders were found to be farmers with a particularly high level of environmental engagement that goes well beyond the more traditional focus on biodiversity to include broader agri-environmental issues such as flood risk management, water quality and aquatic ecology and which require collective management across whole catchments. They are also farmers who tend to run larger farms and be professional farm managers with all that this implies in terms of higher levels of tertiary education, different and more diverse social and professional networks, etc. The career analysis suggests that these farmers are made rather than born, in part as a result of their Environmentally Friendly Farming career, and the role of agri-environment schemes within that.

Environmental Leaders are essential for the 'farmer cluster' approach in which an individual farmer with a specific wildlife interest encourages neighbouring farms to adopt similar conservation management on their own farms in such a way as to benefit wildlife species at the landscape scale. This approach was pioneered by the GWCT and subsequently adopted into Countryside Stewardship nationally. Such clusters require trust, shared objectives and collaboration between neighbouring landowners and result in strong commitments to wildlife conservation over a larger area than is normally possible. They often also involve non-farming members of the local community, strengthening social capital and farmers' pride in what they are doing.

As part of the EU funded SoilCare project, for which the Allerton Project was one of sixteen study sites across Europe, Rust et al. (2018) reviewed the role of social capital in influencing the adoption of sustainable soil management practices. Social capital comprises four core elements – trust, connectedness, social norms and power. Each of these plays an important role. A conclusion from the review was that power was a poorly understood but increasingly influential factor influencing sustainable management as farm sizes and the area owned by absentee landlords increased, field operations were increasingly carried out by contractors, and multinational agricultural companies dominated land management practice. This concern was expressed by our own local farmers, as described above. These farmers also felt disempowered by the global commodity market and by a short-term government agricultural and agri-environmental policy to support them.

> It's almost going to become worse in the future because you're getting fewer, bigger farm units that tend to block crop.
>
> (Water Friendly Farming project farmer)

Trust can be between farmers, or farmers and their advisors such as agronomists, or between farmers and researchers or the farming press. This does not always result in a positive outcome for sustainable management, for example where an agronomist's income is determined by the sale of inputs, especially when products from only one company are offered.

Social norms can become well established, for example where land has been owned by the same family for generations. Some farmers have talked about how they farm partly because it is something they have always done, or because they are carrying on a family tradition. This was not usually conveyed with a sense of obligation but, rather, with a sense of pride in what they do, or simply as something that they had always accepted as part of their life.

> I'm definitely proud to be a farmer, I'm proud to be third generation here and to be carrying that on, the boys will be fourth and that means a lot to me.
>
> (Upper Welland arable farmer interviewed for the SIP)

Through peer pressure, social norms can be either conducive to sustainable land management or contribute to slowing the uptake of new management practices, dependent on whether they are strongly traditional or open to influence from others, as illustrated for the adoption of direct drilling:

> It's still regarded as being rather cranky and rather quirky by mainstream.
>
> (Water Friendly Farming project arable farmer)

Connectedness between farmers is a key aspect of social capital, but it can also be between farmers and other sources of advice or information, or between farmers and non-farming members of the wider community. The relative influence of trust, connectedness, social norms and power differ between farming communities. In our mixed farming landscape, there is an apparent stronger connectedness within livestock or arable sectors than there is between them.

The 2015 SIP survey revealed that 11 of the 34 Welland farmers interviewed had some form of contact with other farmers at least once per week, and 17 did so more frequently. Six farmers had contact with others less than once per week, reflecting the isolation that some farmers experience with current low labour requirements on farms.

> You wave to your neighbour more than you speak to them … one time … you could stop and talk to them but you don't have time anymore.
>
> (SIP survey farmer)

These results were not significantly different from the frequency of contact in the other SIP study sites. Across all study sites, 16% mentioned working with other farmers, and 12% mentioned being involved with some form of a discussion group. Forty-seven percent mentioned some form of social contact such as going to the pub, hunting or shooting together. In a recent comprehensive analysis of gamebird shoots in the UK, Latham-Green et al. (2021) describe how shooting makes a strong contribution to social capital, identity and mental health within rural communities.

Twenty-nine of the 34 Welland farmers had contact with non-farming members of the local community at least once per week, with the nature of these contacts ranging from meeting with friends in the pub, to encountering local residents on footpaths across the farm. As with the farmer to farmer contact, these results were similar across the seven study sites. Across all study sites, 29% were actively involved in community-focused groups or activities such as the village show (14%) or parish council (8%). Twenty percent had some sort of business or farm-related contact with members of the non-farming community. Those who had minimal contact with other members of the community cited the recent increase in the number of incomers, being too busy on the farm, and a lack of village meeting points (e.g. shop or post office).

Nineteen of the 34 Welland farmers had contact with individuals, organisations or companies at least once per week through the sale of farm products, again broadly in line with the other study sites. Nationally, this included livestock markets (35%), agricultural merchants (11%) and direct sales to the public (10%).

From farm to landscape scale

Collaboration between farmers extends the influence of social capital beyond the farm boundary and the level at which it occurs is a major determinant of the extent to which agri-environmental land management can be extended to the landscape scale. Across all farms in the 2015 SIP survey (Morris et al., 2017), the main activities were involvement in trade union (71%), sharing machinery (44%) and labour (39%), discussion groups (37%), involvement in buying groups and producer organisations (both 36%), short-term keep of livestock for or by others (34%), and environmental management (22%). There was a tendency for large farms to be more likely to be involved in environmental management, and very large farms were more likely to be involved in discussion groups.

In the Welland, 19% of farms were involved in machinery sharing, and 9% with contract growing of crops (citing 'economies of scale'), but an additional 22% (the highest across all study sites) were involved in 'other' unspecified collaborative activities). There was a consistent tendency across study sites for a preference for informal agreements over formal ones, although formal agreements were accepted to be necessary for large-scale initiatives involving large sums of money. While formal agreements provide clarity, avoiding disputes, informal ones were regarded as being more flexible and as not being binding.

The main reason given by Welland farmers for cooperating was financial (41%), with others being knowledge exchange (15%) and mutual support (12%). Across all study sites, only 5% mentioned 'environmental' as a reason for collaboration between farms. More farmers mentioned knowledge exchange as a benefit of collaboration than a reason for doing it in the first place, suggesting that it was sometimes an unanticipated consequence of collaboration for other reasons. Five percent of farmers mentioned economic enablers for collaboration, 21% mentioned organisation and governance issues, but a large majority (68%) mentioned social enablers such as trust.

In the 2016 SIP workshops (Fish et al., 2017), farmers viewed themselves as willing and active collaborators and expressed many advantages and examples of farmers collaborating within and between sectors, and across localities. These advantages were expressed primarily in terms of economic benefits for individual farm businesses. Participants often distinguished between the general willingness of farmers to collaborate with each other, and the need for greater partnership working between farmers and wider agencies, academics and government.

Farmers participating in the workshop typically emphasised the need for informal and 'bottom-up' approaches, both as a model of good collaborative practice and a statement of how much collaboration occurs. Ideas of trust were again an important dimension of the success of any arrangements. Participants emphasised the virtues of 'give and take' forms of collaboration epitomised, for instance, by arrangements that avoid money changing hands. Many participants were keen to explain collaboration as a process where farmers routinely looked to each other for insight. Such informal and localised networks of knowledge exchange could be accessed and built quite quickly but extended increasingly to the sharing of information across social media. A local example is provided by a WhatsApp group set up for beef and sheep farmers in 2020. Discussion points have been wide ranging and have included medication, practicalities of fencing, various issues associated with cover crops, comparison of scanning and lambing percentages, the timing of sales, and dung beetles.

Collaboration may be limited by the independent mindset of farmers and the view that farmers had of themselves as being in competition with each other. Participants almost uniformly saw the value of farmers being involved in discussion groups to share ideas and good practice. The attributes of a good discussion group were regarded by SIP workshop participants as being its organised informality, practical setting and involvement of people with similar outlooks and motivations.

Many farmers described machinery sharing as a helpful collaborative activity, but one that was often hampered by a lack of financial capital to invest, varying duties of care towards equipment, a lack of consensus over what machinery to buy, as well as seasonal demands on machinery use.

Participation in agri-environmental schemes and other environmental initiatives, such as Catchment Sensitive Farming (CSF) featured prominently in discussions of collaboration. Farmers emphasised the need for collaboration in these contexts to be dictated by bottom-up processes, with the need for strong independent coordination. There was concern about the congested nature of coordination in some locations and that opportunities were fragmented over a range of funds that farmers sometimes struggled to make sense of. The following exchange in the workshop illustrates the tensions:

FARMER 1: It's better for someone from outside to 'buy in' the community to coordinate that...

FARMER 2: That person could coordinate with the knowledge he would be able to suggest how the collaboration could work ...

FARMER 1: And also how best to get an agreement as farms don't always know the benefits of it short term or long term.

FARMER 3: Need something you can take to the whole area or the whole scheme or whatever don't you, you know to sort of see it from everybody's point of view.

FARMER 4: Yeah, it needs quite a technical eye as well because there's a lot of implications obviously agreeing and things like that, and any changes; it's not a simple hurdle to overcome... But I think ... people have got to be invited to do it, it can't be 'you've got to do this' ...

These comments highlight that there is a need for strong farmer ownership of landscape scale management, with coordination from within the farming community, in combination with technical support and guidance from advisors who are familiar to, and trusted by them. This is especially important given the risks and uncertainties associated with running a farm business. This is in line with the findings from Jarratt (2013) described above. Individual farmers have differing motivations and circumstances which influence their 'environmental careers' but agri-environment schemes play an essential role in supporting them along these career pathways. Such support is especially important given the current economic, political and climate change uncertainty referred to at the start of this chapter.

References

Ang, F., Areal, F., Mortimer, S. & Tiffin, R. (2016) *SIP Project 1: Integrated Farm Management for Improved Economic, Environmental and Societal Performance (LM0201). Work Package 1.1B: Developing Augmented Efficiency and Productivity Measures to Form a Basis for Integrated Sustainable Intensification Metrics.* Defra, London.

Defra (2018) *A Green Future: Our 25 Year Plan to Improve the Environment.* Department for Food and Rural Affairs, London.

Fish, R. (2017). Farmer Discussion Groups – Key Findings. *Report for Defra project LM0302 Sustainable Intensification Research Platform Project 2: Opportunities and Risks for Farming and the Environment at Landscape Scales.* Defra, London.

Hanington, B. (2003). Methods in the making: a perspective on the state of human research in design. *Design Issues* 19 (4) 9–18.

Headey, D. & Fan, S. (2010) Reflections on the global food crisis. How did it happen? How has it hurt? And how can we prevent the next one? *International Food Policy Research Institute. Research Monograph 165.*

Jarratt, S. (2013) *Linking the environmentally friendly farming careers of farmers to their effective delivery of wildlife habitats within the East of England.* Unpublished PhD thesis. University of Nottingham.

Jarratt, S., Morris, C. & Stoate, C. (2013) The role of Environmental Stewardship in the development for farmers' 'environmental learning careers'. Rethinking Agricultural Systems in the UK. *Aspects of Applied Biology* 121, 149–156.

Jones, S. (2019) *A novel farming knowledge cultures approach to the study of soil quality within the context of arable farming in the East Midlands of England.* Unpublished PhD thesis. University of Nottingham.

Latham-Green, T., Hazenberg, R. & Denny, S. (2021) Examining the role of driven-game shooting as a psycho-social resource for older adults in rural areas: a mixed-methods study. *Ageing and Society.* DOI:10.1017/S0144686X2100091X.

Lynch, J., Skirvin, D., Wilson, P. & Ramsden, S. (2018) Integrating the economics and environmental performance of agricultural systems: a demonstration using Farm Business Survey data and Farmscoper. *Science of the Total Environment* 628. DOI:10.1016/j.scitotenv.2018.01.256

Morris, C., Jarrett, J., Lobley, M. & Wheeler, R. (2017). *Baseline Farm Survey – Final Report. Report for Defra project LM0302 Sustainable Intensification Research Platform Project 2: Opportunities and Risks for Farming and the Environment at Landscape Scales.* Defra, London.

Natural England (2012) *Agricultural land classification: protecting the best and most versatile agricultural land. Access to Evidence.* Retrieved 19 December, 2017, from http://publications.naturalengland.org.uk/publication/35012

Rust, N., Noel Ptak, E., Graversgaard, M., Iversen, S., Reed, M., de Vries, J., Ingram, J., Mills, J., Neumann, R., Kjeldsen, C., Muro, M. & Dalgaard, T. (2020) Social capital factors affecting uptake of sustainable soil management practices: a literature review. *Open Emerald Research.* DOI:10.35241/emeraldopenres.13412.2

Stoate. C., Jones, S., Crotty, F., Morris, C. & Seymour, S. (2019). Participatory research approaches to integrating scientific and farmer knowledge of soil to meet multiple objectives in the English East Midlands. *Soil Use and Management* 35 (1) 150–159.

9   A way forward

The Neolithic and Bronze Age settlers at Loddington were probably familiar with the triskele, the three interlocking circles symbol that was adopted across most of Europe to represent and revere the essentials of life – earth, water and air. Agriculture has evolved considerably since then, not least through scientific developments over the past century or so; just 1% of agricultural history. That research has led us to the humbling realisation that we need to respect nature if we are to benefit from it.

Healthy soils, healthy water and healthy air are fundamental requirements for our existence and the quality of each is a reflection, not just of our ability to manage them sustainably, but of the health of our own society. On land and in water, the species we share these resources with can be considered as indicators of the health of the ecosystems on which we all ultimately depend.

During the Covid-19 pandemic, the presence of the virus in water from sewage treatment works was used to monitor the prevalence of different strains in the population (Medema et al., 2020), and for many years, antidepressants have been present in water in sufficiently high concentrations to affect the behaviour and reproduction of a range of aquatic invertebrates (Fang & Ford, 2014). Water bodies with relatively high ecological status are valued by local communities for a range of cultural and recreational uses (Ferrini et al., 2014; Hampson et al., 2017) and their degradation therefore erodes their capacity for supporting mental health.

As discussed in Chapters 6 and 7, high concentrations of phosphorus in the water reflect the inadequacies in our food system, both in terms of the composition of our diet, and in terms of our use of resources in food production. It is important that the response is not just to treat the water, but to address the multiple causes of the problem. That was also a message to come out of our work on sediment and particulate phosphorus in water described in Chapters 3 and 6 – reducing both is more likely to be achieved by improving the management of soil than by creating sediment traps to capture material once it has started moving. We have demonstrated that permeable dams in headwaters reduce downstream flood risk but their ability to do so is reduced as the intensity of storm events increases in line with

DOI: 10.4324/9781003160137-9

climate change expectations. Reducing our contribution to climate change (mitigation) is the priority.

Emissions of nitrous oxide from ruminants grazing grass without access to sufficiently diverse forage (Chapter 7), and from soils that have been compacted in wet conditions by heavy machinery (Chapter 3) provide a reason to modify our current livestock and arable systems. As we found in Chapter 3, soil compaction also has a negative impact on soil ecology and function, including water infiltration rates. Increasing organic matter benefits both, while also sequestering carbon.

There is a need for a network of interstitial semi-natural habitats, outside the productive area but adjacent to it, to support pollinating insects and invertebrate predators of crop pests so that integrated crop management approaches can be adopted (Chapter 5). Vertebrates such as birds also benefit from this integration of semi-natural habitat with the productive area (Chapters 4 and 5). In such managed landscapes, bird species communities themselves may need to be managed, with control of some predatory species in order to maximise bird species diversity or protect certain species (Chapter 7). Equally important is the need to improve management of the productive area through measures such as crop rotations, maintaining green cover and biological control of pests and diseases (Chapter 3). This involves a fine-scale combination of land sparing and sharing to achieve multiple objectives for the farmed environment at the farm and landscape scale.

It is worth noting that, while many species increased in abundance at Loddington in response to the suite of management practices adopted there over the 30 years of the project, some did not. In our work in the Alentejo region of Portugal, we also found that while the abundance of some iconic flagship species such as great bustard *Otis tarda* and little bustard *Tetrax tetrax* increased in areas where targeted management practices were adopted, relative to those where they were not, other species, including some of European conservation concern did not (Santana et al., 2014). Targeted, evidence-based management is necessary for the conservation of some species, and this management can result in increases in others, but it is not inevitable that all species that apparently share the same requirements will benefit to the same extent. For example, as discussed in Chapter 7, landscape scale or regional trends in abundance may influence whether improved productivity or survival results in an increase in abundance on the managed site or dispersal to other areas.

Public payments for public goods

The need to reward farmers for delivering these societal benefits is now widely recognised and accepted. The UK's Environmental Land Management scheme (ELM) is a pioneering initiative that will make public money available to farmers at a range of levels of engagement. As with previous agri-environment schemes, farmers will be rewarded for carrying out management that improves biodiversity, water quality and soil properties, while also contributing to climate change adaptation and mitigation. The intention is to supplement these public payments with private funds such as those through carbon and biodiversity offsetting from industry. This would complement existing initiatives run by food processors and retailers for price premiums for products from farming systems that meet verifiable sustainability criteria (e.g. LEAF Marque, Reed et al., 2017).

Agri-environment scheme payments have been based on the cost of management, plus income forgone from normal agricultural production. This is not a straightforward calculation because, as outlined in Chapter 8, income from sales of farm products varies temporally according to global markets, and costs of management and production vary between farms according to tenure, soil type, topography and farming system. Under the new scenario, farmers are to be paid for delivering public goods and ecosystem services, rather than being compensated for deviating from traditional agricultural practice. However, in order to comply with World Trade Organization 'Green Box' rules, it is anticipated that government payments will focus on costs of management plus income forgone, much as in previous schemes, but with an expectation that additional income for delivery will be secured through alternative mechanisms such as offsetting, agri-tourism, etc. ELM payments will reward farmers for existing environmental management as well as implementing new management practices in order to ensure that those who are currently adopting best practices and pioneering improved environmental management are not penalised.

However, there are substantial flaws in plans for carbon offsetting which need to be addressed if they are not to simply represent a licence to continue net carbon emissions (Allen et al., 2021; Haya et al., 2020). Long-term additionality will be particularly important for mechanisms such as carbon or biodiversity offsetting – for offsetting it will be important to demonstrate that payments are not being made to farmers for management that would have happened anyway, such as the retention of existing landscape features.

As discussed in Chapter 3, carbon sequestered below the plough layer and in a stable rather than labile form is necessary to ensure longevity of any benefit. Establishment of woodland in pasture can reduce rather than increase soil carbon because of suppression of the herb layer and increased microbial activity at depth, although these mechanisms are not fully understood (Douglas et al., 2020; Upson et al., 2016). It is recognised, especially on a global scale, that tree planting can only be part of the solution and can even have negative impacts by displacing other socially and environmentally sustainable land management, and creating a distraction from the overriding need for measures to mitigate climate change at personal, national and global scales (Sen & Dabi, 2021; Waring et al., 2020). However, we have demonstrated clear advantages of introducing trees into agricultural systems in terms of biodiversity and reducing greenhouse gas emissions from ruminants (Chapter 7). Such an approach could be achieved by combining government payments such as those through the ELM with payments from industry and private on-farm initiatives.

It will be essential that each of these funding streams is operational as, alongside the introduction of this new hybrid mechanism, area payments to farmers are being withdrawn. Based on figures for the 2014/2015–2016/2017 period, 53% of lowland grazing farms and 55% of mixed farms are estimated to have made an economic loss in the absence of direct area payments (Defra, 2018). Given the importance of this payment to the economic viability of more than half of farm businesses, it will be important that farmers are adequately rewarded for delivering societal benefits such as flood risk management, carbon sequestration, improved water quality and terrestrial and aquatic biodiversity, etc. Placing a monetary value on these is difficult, or in the case of intrinsic values attached to species at least, virtually impossible (Sandler, 2012). Nevertheless, the conceptual framework of ecosystem services provides a mechanism for valuation of biodiversity in a broader sense, alongside other environmental criteria (Bateman et al., 2013; Dasgupta, 2021).

The new environmental 'market' raises the question of quantification of delivery at a range of scales, but especially at the scale of the individual farm for evaluation of the benefit. This requires monitoring at a range of levels of intensity. Soil carbon can be monitored across a number of sampling points every few years, while an assessment of water quality would require very frequent sampling to capture changes associated with rainfall and vegetation cover etc (Chapters 3 and 6). As discussed in Chapter 6, annual surveying of aquatic invertebrates could be used as a meaningful proxy for the latter but would be expensive and can rarely be linked directly to individual farms. There are also pitfalls associated with soil carbon assessments such as small-scale spatial variation within fields, and analytical differences between laboratories. Spatial mapping data can be used to estimate the soil carbon across large areas but in our analysis of data for Loddington and other UK sites, Beka et al. (2022) have shown that higher resolution spatial data are necessary to provide the most reliable estimates, but these still underestimate soil carbon on high carbon soils. Such disparities would affect payments that farmers receive and introduce uncertainty into future evaluation on the ground.

Grassland plants can be surveyed annually at the appropriate time of year. Birds are also readily surveyed, with well-established techniques being adopted by professionals and volunteers to monitor change in abundance from year to year. However, unlike grassland plants, birds are mobile which introduces inaccuracies at an individual field or farm scale. On a larger scale, many birds are migratory and their breeding abundance is influenced by conditions on migration or in wintering areas (Chapter 7) which are beyond the control of the farmer. Wildlife abundance can also be influenced by other external factors such as climate change. Such external factors also influence the cost of delivering public goods, and more fundamentally, the *ability* to do so and can result in failure to deliver the intended benefits, despite the best endeavours of the farmer (e.g. Walker et al., 2018).

Despite these challenges and uncertainties, one approach that has been explored for rewarding farmers is payment by results, in which payments to farmers are based on the level of delivery of biodiversity (Keenleyside et al., 2014). This can work for the grassland plants example provided above, but less well for others such as birds for the reasons already given. In some cases, it may be necessary to assess the delivery of habitat *quality* such as yield estimates of seed-bearing crops for winter bird food, rather than the numbers of birds themselves (Chaplin et al., 2019). Payment rates can also be established through reverse auctions which result in contracts being awarded to farms delivering the societal benefits for the least cost to the tax-payer or industry funder. However, this approach could severely disadvantage businesses incurring high costs such as tenanted farms, small family farms with disproportionately high transaction costs, or those with low income which, in many cases, are the most appropriate for diversification into environmental management.

WTO Green Box rules preclude, or at least require minimisation of domestic support mechanisms that distort trade by giving a competitive advantage to domestic producers. However, regional, national and international administrations across the world have recognised an environmental emergency and it is important that the potential to address this is not compromised by such restrictions. Examples were given in Chapter 8 of win-wins in which improved nutrient use efficiency or a diverse crop rotation delivered economic as well as environmental benefits, but this was not consistent across farms and was generally marginal for the industry as a whole.

Such synergies are important to the successful and widespread adoption of more sustainable farming systems. They can also contribute to climate change resilience and therefore to food security at the national level. Climate change-induced political instability in the Middle East in 2008 caused food and fertiliser shortage and price inflation. The UK's departure from the EU in 2021 also highlighted the potential of political disruption and labour shortages to reduce food availability in the shops. This issue of national food and fuel security was highlighted still further by the invasion of Ukraine by Russia in February 2022 when fertiliser prices increased three-fold in a few days. Food security is a public benefit that is likely to become an increasingly important consideration as climate change influences global food production and supply and political instability affects the availability of key resources such as fuel and fertiliser. Farming systems that optimise the use of resources at national, regional and even farm scales can contribute to environmental objectives as well as improve the resilience of national food production (Benton et al., 2021).

Current diets in the UK fall short of the recommendations of Public Health England's Eatwell Guide (Scheelbeek et al., 2020) and diets that are recommended by the guide tend to have lower greenhouse gas emissions, as well as having health benefits. An important consideration is that accessible food should be of the required nutritional quality as well as quantity of supply. This is particularly the case for meat for which the nutritional value and climate change impact varies considerably with production system as well as livestock species, with British grass-fed beef and sheep which often support important grassland ecosystems often performing well for both criteria (Barnsley et al., 2021; Mcauliffe et al., 2018).

Climate change adaptation and mitigation

Chapters 6–8 described management practices that address the impacts of climate change on water quality, aquatic biodiversity, flood risk and crop production in agricultural headwaters, and reduce greenhouse gas emissions from grazing livestock. Creation of clean water ponds has very clear biodiversity benefits. It has little impact on the productive land and can enhance the landscape and its inherent interest in a way that is acceptable to farmers. In-channel permeable dams are similarly outside the productive area but the storage of water on adjacent land has the potential to reduce agricultural production, both directly while under water, and indirectly as subsequent waterlogging reduces the period in which the land can be grazed, used for vehicle access, or worked by arable machinery.

There is considerable scope for debate around the payments that might be made to farmers to deliver societal benefits in terms of flood risk management given the uncertainties associated with direct impacts on production and associated indirect impacts. Where payments to farmers are based on expected reductions in downstream flood risk and quantifiable damage to property, that uncertainty increases further. Farmers have also expressed concerns about maintenance costs, and liability for negative consequences that might arise, either on-site or downstream in the case of a dam collapse.

Clean water ponds and permeable dams for flood risk management are clearly adaptation measures providing societal benefits, with clean water ponds also being accepted by farmers as inherent features of their farmland landscape. However, reducing the intensity or frequency of cultivation is both an adaptation and a potential mitigation measure, as described in Chapter 3. Reduced soil disturbance can have advantages in terms of societal environmental, social and economic benefits, while also improving soil function from an agricultural perspective, potentially contributing to enhanced economic performance of farms, but there are also clear barriers to adoption, and potential costs, at least on clay soils.

As described in Chapter 3, there are other benefits associated with reduced soil disturbance through direct drilling or incorporation of grass leys into the rotation, in that there is potential for carbon sequestration in the soil profile (Cooper et al., 2021; Mangalassery et al., 2015). Deep-rooting grasses in grass leys can help to improve water infiltration rates during heavy rainfall, potentially reducing flood risk and improving water quality downstream, but this is likely to be achieved only where shallow soil compaction is not created by intensive autumn grazing of wet soils, or where the sward is not intensively harvested for hay or silage, reducing root growth at depth (Chapter 3). Suppressed root growth will also limit climate change mitigation in the form of sequestration of stable forms of carbon at depth. While there are societal benefits of deep-rooting grass cultivars in the form of both adaptation and mitigation, these are therefore likely to be conditional on constraints on farm businesses, although potential private benefits associated with deep rooting may occasionally arise in drought conditions.

Feeding willow leaves to ruminants reduces greenhouse gas emissions from urine patches and is therefore a clear climate change mitigation measure (Chapter 7). There are also potential modest benefits to farmers in the form of dietary mineral supplementation, especially cobalt and resulting synthesis of vitamin B_{12}. However, this management practice would require substantial changes to the way most livestock farmers currently manage their animals. The need to support farmers' adoption of climate change mitigation measures is considerable, as it is in other sectors such as transport, energy and housing. This is made clear by the reduction in flood peaks achieved by installing permeable dams. There was a 14.5% reduction for 1-in-2-year events; but for 1-in-100-year events which are expected to become more frequent in future, the reduction was just 1% and protection of assets at risk of flooding was therefore minimal, highlighting the importance of climate change mitigation.

The relationships between climate change adaptation and mitigation, and private and public benefits are presented conceptually in Figure 9.1.

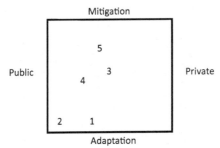

1. Clean water ponds
2. Permeable dams for flood risk management
3. Direct drilling
4. Deep-rooting grasses
5. Willow feeding to ruminants

Figure 9.1 Climate change engagement matrix representing the continuum between climate change mitigation and adaptation, and between public and private benefits. 1. Clean water ponds. 2. Permeable dams for managing downstream flood risk. 3. Multiple objectives for arable soils. 4. Deep-rooting grasses, 5. Willow feeding to ruminants (adapted from Stoate et al., 2022).

Management practices that contribute to climate change mitigation result in strongly societal rather than private benefits and are consequently positioned in the top left of the figure. As such they require some form of economic support to encourage adoption. There is also a need to identify opportunities for funding from individuals or businesses to carry out management practices that deliver private benefits alongside public ones, but public and private interests are not always complementary. At one time, a reluctance to provide state support for soil carbon sequestration because of potential economic benefits to participating farm businesses was a perverse consequence of the public/private dichotomy. There is a need to better understand soil carbon sequestration resulting from different management approaches, and to develop mechanisms that enable farmers to be appropriately rewarded for management that delivers societal benefits. In the specific case of soil organic matter, the water infiltration, flood risk management and water quality benefits associated with it provide additional justification for economic support to deliver these societal benefits.

Farmers' interests and values

As discussed in Chapter 8, Jarratt (2013) revealed how farmers' engagement with biodiversity and its conservation varies considerably between individuals. In terms of agri-environment schemes, farmers' participation can be considered broadly in terms of those for whom income is the primary motivator and those for whom there is an existing interest in wildlife and the funding provides the opportunity to carry out conservation management (although inevitably there is a continuum between these two groups). For the latter, an interest in wildlife can develop into a passion for the conservation of individual species, and even to recruitment of neighbouring farms to collective landscape scale management for one or more species. The cultural attachment to these iconic species can therefore be an important motivator. Land management for gamebird shooting is also often compatible with wider conservation objectives (Chapter 7). These inherent interests within the farming community need to be accommodated in plans for future management, and where possible, can form the foundation for land-use change in which there is an existing commitment from those responsible for it.

As also discussed in Chapter 7, local trusted advisors provide an important catalyst for positive change along the 'environmental careers' of farmers. They are fundamental to the success of such projects as they understand both the diverse interests and circumstances of individual local farmers, and the often complex and evolving nature of regional and national agri-environmental policy. They provide the mechanism for optimising the adoption of a bottom-up approach to agri-environmental management which is more likely to be sustainable in the long term than top-down imposition of environmental policies.

Stoate et al. (2019) evaluated participatory research approaches in which farmers were actively involved to varying degrees in the implementation and execution of research projects concerned with soil management at and around Loddington. Such farmer engagement is essential to practically relevant research although the level and nature of participation might justifiably vary according to the objectives of the research. This approach, whether to research or practical management, ensures that proposed work is practically grounded, and helps to identify values that are common to participating farmers, and how these values are aligned with those of policy makers, regulators and other stakeholders for societal objectives.

In the Welland river basin, through the Welland Valley Partnership, a Resource Protection Group comprises the statutory agencies, regional water company, farming representatives, Welland Rivers Trust, other NGOs and agricultural advisors. The broad range of interests and expertise within the group enables common objectives to be identified and acted upon in the form of farmer advisory visits, dissemination of locally relevant research results, workshops, demonstrations, and other activities. This inclusive approach minimises the risk of conflicts between diverse interests and optimises the use of financial and intellectual resources to achieve real change on the ground at the landscape scale.

Amos (2007) conducted a questionnaire survey of local residents in the upper Eye Brook catchment to explore attitudes to land-use change. They were asked to comment on the merits of continuing the current land use (*status quo*), increasing production of biofuels, improving water quality, and rewilding to benefit wildlife. The biofuels scenario was the least favoured for the community as a whole, with more educated members of the community especially tending towards this opinion. Rewilding was the most favoured at a personal level but ranked in third position for the community as a whole. Men especially preferred the rewilding option. Younger respondents favoured the biofuels scenario more than older ones did. There was considerable disagreement amongst women over and under 55; older women ranked biofuels last and the water quality scenario first, while the reverse was true for younger women. Respondents with farming links favoured the *status quo* and biofuels options. The water quality scenario provided the strongest common ground between farming and non-farming respondents, providing an opportunity for developing a strategy that meets multiple objectives. The survey was a snapshot in time and attitudes may have changed since the survey was conducted. For example, negative aspects of biofuels production came to public attention after the survey was completed. Nevertheless, the results reveal that there are differences in opinions between various members of the local community according to age, gender and involvement in farming.

Landscape scale synergies and trade-offs

We have previously demonstrated that there are both synergies and trade-offs between bird conservation and other ecosystem services (Bradbury et al., 2010), depending on the management adopted and consideration of these interactions is important to achieving multiple objectives as a range of scales. Working in the Welland river basin, and specifically in the Eye Brook tributary, Rayner (2021) developed a methodology to quantify and map the supply of ecosystem services under different scenarios of land-use change. Based on a comparison between 2018 and four future land-use scenarios, he identified trade-offs and synergies between four ecosystem services: (i) water quality, (ii) pollinator abundance, (iii) crop production and (iv) carbon storage.[1]

A fine resolution land-use/land-cover map (LULC) was created from Copernicus Sentinel-2 satellite data with a spatial resolution of 10 m. A cloud-free mosaic of eight images was processed in Google Earth Engine (Gorelick et al., 2017) using a random forest classification algorithm (Breiman, 2001). Five LULC classes were defined: (i) Woodland, (ii) Grassland, (iii) Arable, (iv) Farm Buildings and (v) Suburban.

The land-use change scenarios represented various levels of adoption of ELM policies with a focus on objectives for the aquatic environment. First, areas were selected for potential land-use change based on incremental thresholds of four criteria: flood risk, agricultural land class (ALC) grade, connectivity to the watercourse and proximity to existing areas of high-value biodiversity. In Scenario 1, the most marginal arable areas that are most hydrologically connected to the watercourse were selected for change. Higher-quality arable land of decreasing hydrological connectivity was selected as the thresholds of the criteria were expanded in later scenarios, with buffers adjacent to existing areas of high biodiversity value introduced from Scenario 2, contributing larger areas of semi-natural habitat, ultimately leading to large, connected woodland surrounded by grassland habitats in Scenarios 3 and 4. The area of woodland was adjusted to create an increasing proportion of woodland to grassland as the scenarios progressed (Figure 9.2).

Trade-offs and synergies were assessed for three pairs of ecosystem services: (i) carbon storage and pollinator abundance (Figure 9.3), (ii) carbon storage and phosphate pollution (Figure 9.4) and (iii) crop production and sediment transport to the watercourse (Figure 9.5). There is a well-defined synergy between carbon storage and pollinator abundance (Figure 9.3). Areas where arable land has been converted to grassland or woodland are associated with increased carbon storage but the increase in semi-natural habitat also benefits pollinators. Conversion from arable land to semi-natural habitats also increases pollinator abundance in locations where no land cover change has occurred.

Landcover change between baseline and scenarios

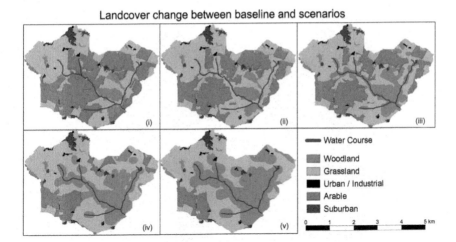

Figure 9.2 The upper Eye Brook catchment study area showing the land-use change scenarios adopted to explore synergies and trade-offs in ecosystem services (Max Rayner, 2021)

Pollinator Abundance vs. Carbon Storage

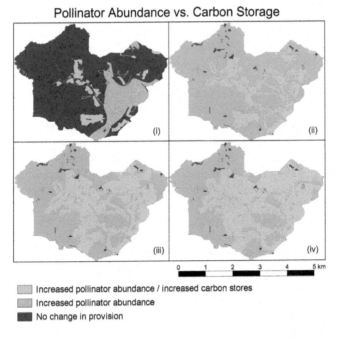

Figure 9.3 Map of the upper Eye Brook catchment study area showing synergies between carbon storage and pollinators (Max Rayner, 2021)

Diffuse Phosphate Pollution vs. Carbon Storage

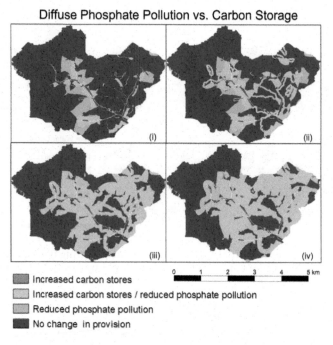

Figure 9.4 Map of the upper Eye Brook catchment study area showing synergies between carbon storage and phosphate pollution reduction (Max
Rayner, 2021)

There is also a synergy between carbon storage and the reduction of phosphate in water (Figure 9.4). Conversion of arable land to semi-natural habitats close to the watercourse increases carbon storage and reduces phosphate
pollution, and phosphate pollution of water from areas upslope of these regions where no land cover change has occurred is also reduced (Figure 9.4 (i)
and (ii)). The widespread land-use change associated with Scenarios 3 and
4 leads to extensive areas of synergy between carbon storage and reduced
phosphate pollution (Figure 9.4 (iii) and (iv)).

Unlike the previous two pairings of ecosystem services, there is a trade-
off associated with crop production and the conversion of arable land to
semi-natural habitats to reduce sediment transport to the watercourse
(Figure 9.5). Crop production is obviously reduced as arable areas are converted to semi-natural habitats, but this change is associated with reduced
sediment transport in the areas where crops have been removed, and also
in the upslope areas where crops remain in place but converted downslope
areas act as a buffer. As more areas are converted from arable land to
semi-natural habitats, the extent of this trade-off increases.

The dynamics of ecosystem service supply associated with the land-use
scenarios can also be quantified in terms of the aggregate areas of ecosystem

Sediment Transport vs. Crop Production

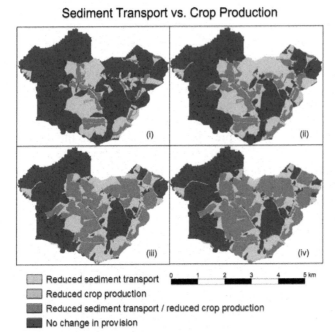

Reduced sediment transport
Reduced crop production
Reduced sediment transport / reduced crop production
No change in provision

0 1 2 3 4 5 km

Figure 9.5 Map of the upper Eye Brook catchment study area showing trade-offs between food production and reduction in sediment transport to water (Max Rayner, 2021)

service provision. For the first two pairings of ecosystem services the benefits of the scenarios (in terms of synergies) only increase as land-use change becomes more extensive. Greater areas of the catchment can provide increased provision of the targeted ecosystem services. However, for crop production and sediment transport to water, the situation is not so simple. This becomes apparent when comparing ecosystem service provision between Scenario 2 and 3. The total area from which sediment transport to water is reduced in Scenario 2 is 474 ha which is associated with reduced crop production across 166 ha of the catchment. However, in Scenario 3 the total area from which sediment transport to water is reduced increases to 570 ha, while reduced crop production occurs across 328 ha of the catchment. Such relationships in the trade-off of ecosystem services associated with land-use change pose important questions to policy makers, specifically in this case – are the gains in reduced soil loss to water between Scenarios 2 and 3 worth the massively reduced levels of food production?

The dynamics of ecosystem service provision modelled seem to accurately represent the physical processes that would be expected from the changes to land use associated with the scenarios. For example, the riparian buffer that is established reduces phosphate pollution and sediment transport from upslope areas where land use is yet to change due to the intercepting effects of the semi-natural habitats that have been established in proximity to the watercourse. Also, the benefits of semi-natural habitat to pollinator abundance are not restricted solely to regions where land use has changed. Pollinating insects will forage for some distance from the locations in which they nest, so the increase in semi-natural habitats has beneficial effects on pollinator populations some distance from the areas of land-use change, spreading the effects across the catchment.

Analyses such as those presented here can provide the data to support these decisions and highlight areas across landscapes where land-use change is associated with synergies and trade-offs in ecosystem service supply. Balancing the needs and views of multiple stakeholders will be one of the greatest challenges facing future land-use policy. However, basing decisions on evidence in the form of data from spatial ecosystem service models can potentially avoid inefficient or detrimental land-use change, ensuring that public money is spent wisely, and that future environmental objectives can be met.

While apparently a top-down policy-driven approach to local stakeholder engagement, the mapping approach can also be used to enable farmers and other individuals to challenge and express their priorities for land management. It is important that such tools should be regarded as a first step in an iterative process of engagement and co-development, with farmers and other land managers having a major role in this process. In our example, the land-use scenarios that guide the mapping of ecosystem services are founded on objectives for catchment scale improvement of the aquatic environment, primarily through taking arable land out of production. However, other approaches such as reduced tillage or direct drilling, short-term fallows, grass leys and cover crops could be adopted instead, or as well as, taking land out of production, as evidenced by our own research described in Chapters 3 and 6. Completely different objectives identified by farmers such as opportunities presented by economically marginal farms or local interest in specific environmental issues or individual iconic species should carry equal weight in developing alternative maps for further discussion. As discussed earlier, such an approach ensures ownership from those responsible for delivering the agreed objectives, increasing the likelihood of successful outcomes. It can be combined with small-scale farmer-led trials that provide a focus for individual and collective learning.

Participatory research on various aspects of soil management has ensured that the research is grounded in local cultural values, is aligned with the economic objectives of farmers, and is relevant to the farming systems adopted by them (Stoate et al., 2019; Figure 9.6). Such a collaborative approach involving researchers and farmers is essential to successful research and the development of future policy, and farmer engagement plays a key role in the piloting of approaches being considered for England's ELM scheme. We

Figure 9.6 Farmers participating in soil management research in the Eye Brook catchment

have carried out participatory research with farmers on the Senegal/Gambia border where tree density had been greatly reduced by clearance for crops and grazing by livestock, soil erosion had increased, and land and adjacent coastal fisheries were degraded (Stoate & Jarju, 2008; Figures 9.7). Applying leaf mulch from some tree species improved crop performance in a simple groundnut and millet rotation. This collaborative research with farmers increased the value they attached to tree species such as *Faidherbia albida* and *Guiera senegalensis* which were subsequently encouraged to regenerate in arable fields, improving the productivity and resilience of the farming system and local food security, while also providing habitat for migratory warblers and multiple non-migratory species. The approach adopted tapped into the local cultural values attached to various tree species as well as addressing food production and security issues that were of immediate concern to the local farming community (Stoate et al., 2001).

Thinking globally

While active involvement of farmers in decision making is important at the local level, climate change and other global factors have an important influence on the ability to carry out the management successfully, and the likelihood of a successful outcome, either through impacts on local conditions, or influences from other parts of the world. Climate change has influenced the distribution of many species, including colonisation of the farm at Loddington by lesser-marsh grasshopper *Chorthippus Albomarginatus* and tree bumblebee *Bombus hypnorum* as part of the north-westwards

Figure 9.7 Farmers participating in soil management research close to the Senegal/ Gambia border

range expansion of these species for example. As discussed in Chapter 7, conditions on migratory birds' wintering grounds, or on their migration, are substantially affected by climate change and can influence the survival and subsequent breeding abundance, independently of conditions on the breeding grounds. Climate change can therefore have positive and negative impacts on the abundance of different species at any particular location.

Climate change impacts in the wintering areas of migratory birds also affect food production in those regions, accentuating the conflict between wildlife and the production of food (e.g. Wymenga & Zwarts, 2010). As described in Chapter 7, destruction of fruiting bushes by cattle and other livestock in the riparian zone of the river Senegal can reduce berry abundance and fat accumulation by migratory warblers, reducing their probability of survival during the spring migration. The insect food of migratory warblers is more abundant in undisturbed woodland with climbing plants in the trees than in woodland where grazing precludes the establishment of these climbing plants. Such habitats become increasingly scarce as climate change and agricultural pressures increase, with both sedentary and migratory species being negatively affected.

Importation of food into the UK improves national food security by diversifying sources of supply, but also threatens it as a result of climate change and political instability in producer countries and puts additional pressure on land use and food production in those countries. A global approach is therefore necessary for wildlife conservation, climate change and food security. While often regarded as independent objectives, all of these are closely integrated.

Food production standards in the UK contribute to meeting targets for nature recovery, climate change and animal welfare. However, there is a consequential impact on domestic food production as a result of increased production costs, reduced yields, and land taken out of production. Food imported to meet domestic demand often does not comply with the same environmental or welfare standards. The result is that UK farmers are unable to compete, and environmental problems are simply exported to other parts of the world. An example is provided by vegetable oils following the ban on use of neonicotinoid insecticides to protect pollinating insects in 2018. As described in Chapter 8, the prevalence of severe damage from cabbage stem flea beetle *Psylliodes chrysocephala* resulted in many UK farmers largely abandoning oilseed rape as a viable component of their crop rotations, and rapeseed and other vegetable oils with lower environmental standards were imported to meet domestic demand. This included rapeseed oil grown with the use of neonicotinoids in other countries (Cunningham & Allen-Stevens, 2020). As rapeseed is also a source of high protein animal feed, lower domestic production is compensated for by the import of alternative products such as soya which are grown to lower environmental standards and associated with large-scale destruction of forest in South America for example. Such international transportation of food and feed also contributes 3% of UK greenhouse gasemissions and this proportion is expected to rise substantially if considerable efforts are not made to maintain UK production (CCC, 2020).

Superimposed on these nuanced implications of international trade for the agricultural environment across the world are the major overriding issues of human population growth, excessive consumption and food waste. The global population is estimated to grow from 7.8 billion today to around 9.7 billion by 2050, ultimately reaching around 11 billion (UN DESA, 2019). As a result of cheaper food that is increasingly energy-rich rather than nutrient-dense, the proportion of the global population that is obese now exceeds those who are malnourished, although the latter remains a large proportion as a result of inequitable distribution (Benton & Bailey, 2019). Around a third of edible food produced for human consumption globally is wasted at some stage along the food chain. If food consumption that is in excess of individual nutritional requirements is included, it has been estimated that 48% of harvested crops are wasted (Alexander et al., 2017). Added to the negative impacts on global biodiversity and climate change, these considerations make a strong case for a major revision of the global food production system.

Thinking locally

The UK food production to supply ratio for indigenous food is currently 76%, indicating considerable potential for increasing home-grown production, especially for fruit and vegetables which are predominantly imported

(Defra, 2018). Increasing production of nutrient-rich food within the UK, including sustainably produced meat as well as fruit, vegetables and low input, high protein pulses would reduce reliance on imports and negative impacts elsewhere. However, since the UK left the EU in 2021, trade deals established with other countries have increased the likelihood that food will be imported from farming systems with lower environmental standards and threatened the economic viability of UK farming systems that adopt high environmental and animal welfare standards.

The example of phosphorus from Chapters 3 and 6 provides an illustration of the need for a localised and internalised food production system that is compatible with meeting multiple environmental objectives. Phosphorus is essential for food production, a finite resource that is mined in politically unstable parts of the world such as Western Sahara and Belarus, and is used inefficiently, causing widespread wastage and eutrophication, especially of the aquatic environment. There is a need to improve phosphorus use on farm, for example through the use of improved soil ecology and, possibly, cover crops to exploit legacy P within the soil. Better integration of livestock and arable systems would provide an opportunity for more targeted application of manures to arable crops, while also introducing grass leys into arable rotations, further improving soil structure and function.

There is also a need to recycle phosphorus more efficiently within the human food system through a reduction in the loss of phosphorus to water from sewage treatment works and more targeted and efficient application of sewage sludge and anaerobic digestate to land. Food waste should be valued as a source of phosphorus, and consumers need to understand the contribution different foodstuffs, especially highly processed foods, make to phosphorus loss from the food system to the environment. Phosphorus therefore provides a classic example of the need for a circular local economy in which food security is strengthened (through less reliance on imports of food or fertiliser, and more sustainable production) and environmental impacts are reduced (through reduced nutrient loss to water). However, for the forthcoming decades, there is expected to be a continuing loss of historically applied P from farmland to water. In this sense, there is a parallel with climate change mitigation. Because of the long time lag between action and response, it is all the more urgent to act now and to do so in an integrated way that avoids perverse outcomes and optimises synergies. Landscape scale initiatives such as the Catchment Based Approach (CaBA) bring together a range of stakeholders and give those managing the land a voice, support, access to advice and training, as exemplified locally by the Welland Valley Partnership.

In Chapter 8, farmers expressed concern that farm size would increase as smaller businesses became unviable, with an increasing area of land being managed by contractors. Given the importance of tenure in influencing long-term objectives for sustainability, such an outcome from the current agricultural transition away from EU policies would reduce the connection between the land and those managing it. What is needed is a knowledge-based system with reduced reliance on external inputs and more efficient use of those that are adopted through a more nuanced understanding of site-specific processes and resources.

As more integrated approaches to pest, weed and disease control are adopted and pesticide use is reduced to minimise resistance and ecological damage, agrochemical manufacturers and suppliers have an important role to play in disseminating locally relevant, evidence-based advice to farmers and developing and supporting provision of appropriate technology for resource use efficiency. Technical solutions can often help to achieve agri-environmental objectives, and continuing development and application of scientific knowledge will be important to sustainable food production in future. But equally important is the local knowledge of farmers on the ground who have an intimate knowledge and attachment to the land they manage. Increasingly, as discussed previously, farmers understand the abundance, distribution, and often the ecology of wildlife species on their land, as well as the idiosyncrasies of their soils. Such understanding is associated with a strong, often long, but always secure connection to the land they farm.

War in Europe in the 1940s highlighted the need to improve national food security, but this was to the detriment of many environmental processes on which food production and much else depends. Russia's invasion of Ukraine in 2022 once again puts the spotlight on national food production, but our understanding of the interdependence between this and environmental objectives is considerably improved, as described throughout this book. In its first 30 years of research, the Allerton Project has demonstrated the integration of technical and practical knowledge and understanding. It highlights the need to operate at a range of scales and to understand and accept often complex issues associated with global food production, trade and ecology, and use this knowledge to develop economically and environmentally sustainable management approaches at the local level. These approaches need to have farmers at their core, but also to reconnect consumers with their food and its ecology.

Notes

1 Ecosystem service provision was estimated using the *Integrated Valuation of Ecosystem Services and Trade-offs* (InVEST) toolkit (Sharp et al., 2020), a series of ecosystem service models that simulate biophysical processes. The models were calibrated and validated with data from the catchment and run for the 2018 baseline, and the four scenarios of potential future land use. Model outputs were converted to identify areas of potential trade-off and synergy. First, using the raster calculator function of QGIS v3.6.1 the baseline and scenario rasters for each service were compared with one another to produce three new binary rasters for each baseline/scenario pairing: 'Scenario < Baseline', 'Scenario > Baseline', and 'Scenario = Baseline'. These three rasters were then converted (so that each had a unique value) and added together to produce a single raster for each baseline and scenario pairing of the four ecosystem services detailing where ecosystem service provision increased, decreased, or remained constant. The 'second step' rasters were then combined for pairs of ecosystem services to establish locations where land-use change created areas of trade-off or synergy.

References

Alexander, P., Brown, C., Arneth, A., Finnigan, J., Moran, D. & Rounsevell, M. (2017) Losses, inefficiencies and waste in the global food system. *Agricultural Systems* 153, 190–200.

Allen, M., Tanaka, K., Macey, A., Cain, M., Jenkins, S., Lynch, J. & Smith, M. (2021) Ensuring that offsets and other internationally transferred mitigation outcomes contribute effectively to limiting global warming. *Environmental Research Letters* 16, 074009. DOI:10.1088/1748–9326/abfcf9

Amos, M. (2007) *Can landscape mapping be used as a tool to engage rural communities and stakeholders in environmental best practice schemes?* Unpublished MSc thesis. Imperial College, London.

Barnsley, J. E., Chandrakumar, C., Gonzalez-Fischer, C., Eme, P. E., Bourke, B. E. P., Smith, N. W., Dave, L. A., McNabb, W. C., Clark, H., Frame, D. J., et al. (2021) Lifetime climate impacts of diet transitions: a novel climate change accounting perspective. *Sustainability* 13, 5568. DOI:10.3390/su13105568

Bateman, I. J., Harwood, A. R., Mace, G. M., Watson, R. T., Abson, D. J., Andrews, B., Binner, A., Crowe, A., Day, B. H. & Dugdale, S. (2013) Bringing ecosystem services into economic decision-making: land use in the United Kingdom. *Science* 341 (6141) 45–50.

Beka, S., Burgess, P.J., Corstanje, R. & Stoate, C. (2022) Spatial modelling approach and accounting method affects soil carbon estimates and derived farm-scale carbon payments. *Science of the Total Environment* 827. doi.org/10.1016/j.scitotenv.2022.154164.

Benton, T. G. & Bailey, R. (2019) The paradox of productivity: agricultural productivity promotes food system inefficiency. *Global Sustainability* 2 (e6), 1–8. DOI:10.1017/sus.2019.3

Benton, T. G., Bieg, C., Harwatt, H., Pudasaini, R. & Wellesley, L. (2021) *Food system impacts on biodiversity loss – three levers for food transformation in support of nature.* The Royal Institute- of International Affairs, Chatham House, London.

Bradbury, R. B., Stoate, C. & Tallowin, J. (2010) Lowland farmland bird conservation in the context of wider ecosystem service delivery. *Journal of Applied Ecology* 47, 986–993.

Breiman, L. (2001) Random forests. *Machine Learning* 45(1) 5–32.

CCC (2020) *The Sixth Carbon Budget – Shipping.* Climate Change Committee, London.

Chaplin, S., Robinson, V., LePage, A., Keep, H., Le Cocq, J., Ward, D., Hicks, D. & Scholz, E. (2019). *Pilot Results Based Payment Approaches for Agri-environment Schemes in Arable and Upland Grassland Systems in England. Final Report to the European Commission.* Natural England and Yorkshire Dales National Park Authority.Cunningham, C. & Allen-Stevens, T. (2020) OSR assurance – the break crop at breaking point? *Crop Production Magazine* March 2020, 20–22.

Dasgupta, P. (2021) *The Economics of Biodiversity: The Dasgupta Review.* Abridged Version. HM Treasury, London.

Defra (2018) *The Future Farming and Environment Evidence Compendium.* Department for Food and Rural Affairs, London.

Douglas, G., MacKay, A., Vibart, R., Dodd, M., McIvor, I. & McKenzie, C. (2020) Soil carbon stocks under grazed pasture and pasture-tree systems. *Science of the Total Environment* 715, 136910. DOI:10.1016/j.scitotenv.2020.136910

Fang, P. P. & Ford, A. T. (2014) The biological effects of antidepressants on the molluscs and crustaceans: A review. *Aquatic Toxicology* 151, 4–13. DOI:10.1016/j. aquatox.2013.12.003

Ferrini, S., Schaafsma, M. & Bateman, I. (2014) Revealed and stated preference valuation and transfer: A within-sample comparison of water quality improvement values. *Water Resources Research* 50. DOI:10.1002/2013WR014905

Gorelick, N., Hancher, M., Dixon, M., Ilyushchenko, S., Thau, D. & Moore, R. (2017) Google Earth engine: planetary-scale geospatial analysis for everyone. *Remote Sensing of Environment* 202, 18–27.

Hampson, D., Ferrini, S., Rigby, D. & Bateman, I. (2017) River water quality: who cares, how much and why? *Water* 9, 621. DOI:10.3390/w9080621

Haya, B., Cullenward, D., Strong, A. L., Grubert, E., Heimayr, R., Sivas, D. A. & Wara, M. (2020) Managing uncertainty in carbon offsets: insights from California's standardised approach. *Climate Policy* 20 (9) 1112–1126. DOI:10.1080/14693 062.2020.1781035

Jarratt, S. (2013) *Linking the environmentally friendly farming careers of farmers to their effective delivery of wildlife habitats within the East of England.* Unpublished PhD thesis. University of Nottingham.

Keenleyside, C., Radley, G., Tucker, G., Underwood, E., Hart, K., Allen, B. & Menadue, H. (2014) *Results-based Payments for Biodiversity Guidance Handbook: Designing and Implementing Results-Based Agri-environment Schemes 2014– 20.* Prepared for the European Commission, DG Environment, Contract No ENV.B.2/ETU/2013/0046, Institute for European Environmental Policy, London.

Mangalassery, S., Mooney, S., Sparkes, D., Fraser, W. & Sjogersten, S. (2015) Impacts of zero tillage on soil enzyme activities, microbial characteristics and organic matter functional chemistry in temperate soils. *European Journal of Soil Biology* 68, 9–17.

Mcauliffe, G. A., Takahashi, T. & Lee, M. R. F. (2018) Framework for life cycle assessment of livestock production systems to account for the nutritional quality of final products. *Food and Energy Security.* DOI:10.1002/fes3.143

Medema, G., Heijnen, L., Elsinga, G., Italiaander, R. & Brouwer, A. (2020) Presence of SARS-Coronavirus-2 RNA in sewage and correlation with reported Covid-19 prevalence in the early stage of the epidemic in the Netherlands. *Environmental Science & Technology Letters* 7, 511–516. DOI:10.1021/acs.estlett.0c00357

Rayner, M., Balzter, H., Jones, L., Whelan, M. & Stoate, C. (2021) Effects of improved land-cover mapping on predicted ecosystem service outcomes in a lowland river catchment. *Ecological Indicators* 133, 108463. DOI:10.1016/j.ecolind.2021.108463

Reed, M., Lewis, N. & Dwyer, J. (2017) *The effect and impact of LEAF Marque in the delivery of more sustainable farming: a study to understand the added value to farmers.* Countryside and Community Research Institute, Gloucester.

Sandler, R. (2012) Intrinsic value, ecology, and conservation. *Nature Education Knowledge* 3 (10) 4.

Santana, J., Reino, L., Stoate, C., Borralho, R., Rio Carvalho, C., Schindler, S., Moreira, F., Bugalho, M., Ribeiro, P., Santos, J., Vaz, A., Morgado, R., Porto, M. & Beja, P. (2014) Mixed effects of long-term conservation investment in natural 2000 farmland. *Conservation Letters* 7 (5) 467–477.

Scheelbeek, P., Green, R., et al. (2020) Health impacts and environmental footprints of diets that meet the Eatwell Guide recommendations: analyses of multiple UK studies. *BMJ Open* e037554. DOI:10.1136/bmjopen-2020-037554

Sen, A. & Dabi, N. (2021) *Tightening the Net: Net Zero Climate Change Targets – Implications for Land and Food Equity.* Oxfam, Oxford.

Sharp, R., Douglass, J., Wolny, S., Arkema, K., Bernhardt, J., Bierbower, W., Chaumont, N., Denu, D., Fisher, D., Glowinski, K., Griffin, R., Guannel, G., Guerry, A., Johnson, J., Hamel, P., Kennedy, C., Kim, C. K., Lacayo, M., Lonsdorf, E., Mandle, L., Rogers, L., Silver, J., Toft, J., Verutes, G., Vogl, A. L., Wood, S. & Wyatt, K. (2020) *InVEST 3.9.0 User's Guide.* The Natural Capital Project, Stanford University, University of Minnesota, The Nature Conservancy and World Wildlife Fund.

Stoate, C., Biggs, J., Williams, P. & Brown, C. (2022) An assessment of the practical potential and level of participatory research needed to meet catchment scale climate change objectives. *Farming Systems Facing Climate Change and Resource Challenges.* 14th European IFSA Symposium, Evora, Portugal. 349–358.

Stoate, C., Morris, R. M. & Wilson, J. D. (2001) Cultural ecology of Whitethroat (*Sylvia communis*) habitat management by farmers: trees and shrubs in Senegambia in winter. *Journal of Environmental Management* 62, 343–356.

Stoate, C. & Jarju, A. K. (2008) A participatory investigation into multifunctional benefits of indigenous trees in West African savanna farmland. *International Journal of Agricultural Sustainability* 6, 122–132.

Stoate, C., Jones, S., Crotty, F., Morris, C. & Seymour, S. (2019). Participatory research approaches to integrating scientific and farmer knowledge of soil to meet multiple objectives in the English East Midlands. *Soil Use and Management* 35 (1) 150–159.

UN DESA (2019) World Population Prospects 2019: Highlights. United Nations, Department of Economic and Social Affairs. ST/ESA/SER.A/423.

Upson, M. A., Burgess, P. J. & Morison, J. I. L. (2016) Soil carbon changes after establishing woodland and agroforestry trees in grazed pasture. *Geoderma* 283, 10–20. DOI:10.1016/j.geoderma.2016.07.002

Walker, L. K., Morris, A. J., Cristinacce, A., Dadam, D., Grice, P. V. & Peach, W. J. (2018) Effects of higher-tier agri-environment scheme on the abundance of priority Farmland birds. *Animal Conservation* 21, 199–200. DOI:10.1111/acv.12386

Waring, B., Neumann, M., Prentice, I. C., Adams, M., Smith, P. & Siegert, M. (2020) Forests and decarbonization – roles of natural and planted forests. *Frontiers in Forests and Global Change* 3, 58. DOI:10.3389/ffgc.2020.00058

Wymenga, E. & Zwarts, L. (2010) Use of rice fields by birds in West Africa. *Waterbirds* 33, 97–104.

Index

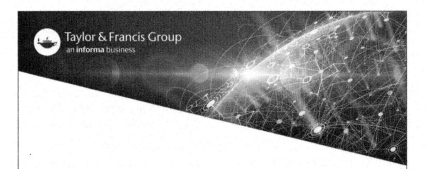